MW00861922

THE SOCIAL GENOME

ALSO BY DALTON CONLEY

You May Ask Yourself:
An Introduction to Thinking Like a Sociologist

Parentology:
Everything You Wanted to Know about the Science
of Raising Children but Were Too Exhausted to Ask

The Genome Factor:
What the Social Genomics Revolution Reveals
about Ourselves, Our History, and the Future
(with Jason Fletcher)

The Pecking Order:
Which Siblings Succeed and Why

Honky

THE
SOCIAL
GENOME

The New Science of
Nature and Nurture

Dalton Conley

W. W. NORTON & COMPANY

Independent Publishers Since 1923

Copyright © 2025 by Dalton Conley

All rights reserved
Printed in the United States of America
First Edition

For information about permission to reproduce selections from this book, write to
Permissions, W. W. Norton & Company, Inc., 500 Fifth Avenue, New York, NY 10110

For information about special discounts for bulk purchases, please contact
W. W. Norton Special Sales at specialsales@wwnorton.com or 800-233-4830

Manufacturing by Lake Book Manufacturing
Book design by Daniel Lagin
Production manager: Julia Druskin

ISBN 978-1-324-09263-6

W. W. Norton & Company, Inc., 500 Fifth Avenue, New York, NY 10110
www.wwnorton.com

W. W. Norton & Company Ltd., 15 Carlisle Street, London W1D 3BS

1 2 3 4 5 6 7 8 9 0

For Tren, a true force of nature

"Give me a dozen healthy infants, well-formed, and my own specified world to bring them up in and I'll guarantee to take any one at random and train him to become any type of specialist I might select—doctor, lawyer, artist, merchant-chief, and yes, even beggar-man and thief, regardless of his talents, penchants, tendencies, abilities, vocations, and race of his ancestors."

—John B. Watson, *Behaviorism* (1925)

"If someone's liver doesn't work, we blame it on the genes; if someone's brain doesn't work properly, we blame the school. It's actually more humane to think of the condition as genetic. For instance, you don't want to say that someone is born unpleasant, but sometimes that might be true . . . We're not all equal, it's simply not true. That isn't science."

—The other Watson
(James D. Watson, he of DNA fame)

TABLE OF CONTENTS

THE SOCIAL GENOME

1

Welcome to the
Sociogenomics Revolution

Nine years ago, my wife and I sat in a fertility clinic, discussing our options with a specialist. It was her first marriage, but it was my second. I had already had two children from that prior relationship who were, by then, teenagers. Fourteen years had passed since the vasectomy I had undergone after my second child was born. Back when I had two kids under the age of three, I had been fairly certain I was done, but now I had fallen in love and wanted to expand my family. The doctor told us that after a decade, vasectomy reversals rarely worked and, moreover, we wouldn't even know if the surgery had been successful for almost two years. That seemed like an eternity at the time.

My first marriage had been way too rocky to handle a third child, but I also recognized that none of us can predict the future. So, before I went under the knife, I stored some sperm in a cryogenic laboratory.

But seven years later, during a quiet period before the final storm of my relationship, I let miserliness get the best of me and filled out the paperwork to dispose of the samples. The storage fees had risen every year, and at the time, it seemed silly to keep throwing money away. I was done having children. Those were sperm from a thirty-two-year-old I had thrown away. Now, I was ruing that decision. I was well aware of the research showing a correlation between paternal age and risk of psychopathology in offspring. The older the dad, the greater chance the kid might suffer from autism, ADHD, or schizophrenia—just to name a few of the adverse possibilities. To take autism as an example, one Israeli study found that men in their forties had a six-fold increase in the chances of having an autistic child than men under thirty.[1], * Another examination of Swedish medical records obtained results that were not as dramatic: a 75 percent increase for men in their late forties (as I was at the time).[2] I had more than one acquaintance who had become a father later in life and had then raised an autistic child, and I saw how hard life was for them.

The baseline percentages for autism were low—around 1 percent—so even a five-fold increase would result in only a 5 percent chance. But one-in-twenty still frightened me, especially when added to all the other possible risks. One of my other children had been diagnosed with ADHD, so that was probably the greater likelihood. It was daunting enough to me to think about starting all over again with a new baby at my age, but to deal with a kid who had some additional challenges, like autism, ADHD, or even schizophrenia, was beyond what I thought I could handle. This was my biggest fear as I stepped into the river of fatherhood for a third time.

Thanks to my frugality, instead of a thirty-two-year-old's sperm, we had to resort to a new batch that would be extracted from my testes and injected, one by one, into my wife's ova. Since we were

---

* A note on how to read this book: There are 173 endnotes. Many of these are simply references to the research studies. Others provide extended, often tangential, discussions on topics that arise in the main text: the debate over genes and free will, why CRISPR is a sideshow, how related siblings actually are, and more. You can safely ignore the endnotes. But if you crave more detail, dipping into them should satisfy that need.

using older sperm now, I wanted to have a girl, who would be at less risk for autism and schizophrenia, which are both more common among males. When we inquired about selecting the sex, the doctor explained that the sex chromosomes would be revealed during pre-implantation genetic testing to assay which embryos were viable and didn't show signs of abnormalities like trisomy-21 (Down syndrome). She added that there was no reason why we couldn't choose based on sex among the healthy embryos. Then I asked another question: "Can we genotype the DNA of the embryos and calculate polygenic indices for them?"[3] In many other countries, even sex selection is illegal except in special cases. In the United States, fertility medicine is a wild west where almost anything goes.

She looked at me quizzically. I explained what a *polygenic index (PGI)* was: a single number that summarized a person's genetic tendency for a particular disease or trait—for instance, blood pressure, height, or cognitive functioning. Whether we know it or not, we all have polygenic indices for hundreds of outcomes: depression, body mass index (BMI), diabetes, and educational attainment, to name a few. And yes, for autism. That number—or rather, those numbers, since there is a different one for each trait—have become the FICO scores of human genetics.

WHEN THE DRAFT SEQUENCE OF THE HUMAN GENOME PROJECT WAS completed in 2003, seventy-five-year-old James Watson—the co-discoverer of DNA—dreamt of soon finding the fundamental causes of all human outcomes in its string of letters. Indeed, scientists thought it would be just a few short years before the five genes behind heart disease, the dozen or so implicated in schizophrenia, and the twenty involved in cognitive ability would be discovered in this so-called dictionary of life. This optimism was well founded. The latest medical genetics had already delivered the genes that were the most common causes of intellectual disability (FMR1), of breast cancer (BRCA1 and BRCA2), and of cognitive decline due to Alzheimer's (APOE4). Once we knew the genes that lay behind the plaque in our arteries, various neurological and blood disorders, and even our tendency to smoke or drink too much, these conditions would become relics of the past. As it turned out, it took a decade to learn that most traits and diseases were influenced not by a handful of genetic differences, but by thousands. Most traits or diseases were highly *polygenic.*

At first, geneticists despaired: How could they understand the fundamental biology of, say, high blood pressure when it involved a multitude of genes? Moreover, what hope was there of ever devising a cure for schizophrenia if there wasn't one, or even a dozen, but rather a thousand genes that a pharmaceutical needed to mimic or block in order to treat the disease—especially when those thousand genes were implicated in many other biological processes in the body? There would not be just a couple of proteins that were the keystones to combatting heart disease or memory decline, proteins whose actions could be blocked (or enhanced) by drugs or, today, by mRNA vaccines. There would be no simple pill to eliminate diabetes. No obvious gene therapy for depression or for cognitive enhancement, for that matter. The 2011 Bradley Cooper film, *Limitless*, about a struggling writer who takes a new drug and ends up a financial wizard, would remain science fiction for the foreseeable future.

Without going too deep into technical terminology: we all have three billion base pairs in our genome. These base pairs are nucleotides that exclusively bond to each other and are often referenced by the first letters of their names: (A)denine, (C)ytosine, (G)uanine, and (T)hymine. A always bonds with T, and C goes with G—hence the term *base pairs.* Only about 0.1 percent, or one in a thousand beads on that string of DNA, differs across people. Hence the common adage that we are all 99.9 percent identical. Understanding how variation in that 0.1 percent affects who we become required measuring DNA bead by bead across all three million strung on our twenty-three pairs of chromosomes.

The good news was that once cheap, genome-wide data became available for large numbers of people, we could test almost all those beads, and not just a handful we suspected might be associated with disease, using a method called a *genome-wide association study* or GWAS (pronounced "g-wass"). The first GWAS was run in 2005 on the now puny sample size of 96 cases of people with macular degeneration along with 50 controls. With just those 146 subjects, the authors of that pioneering study were still able to locate a key gene that increased risk for this eye disease by a factor of over seven-fold.

Another early and important GWAS was conducted on schizophrenia—a condition that bedevils about 1 percent of the population; its symptoms are devastating to sufferers (and their families). The typical schizophrenic lives twelve to fifteen years less than a non-sufferer. And the quality of those years is much reduced, as anyone with a relative or friend who has the disease well knows. Schizophrenia's onset in early adulthood is difficult to predict. But scientists have known for a long time that it is highly heritable—that is, that the likelihood of getting it is influenced by one's genetic makeup. Most twin and adoption studies had arrived at a figure of 80 percent for its heritability, meaning that four-fifths of the variation in incidence in a population is due to genetic differences within that population. Some scholars thought that schizophrenia resulted from rare mutations that had big effects; others posited that it resulted from thousands of tiny effects across all the chromosomes.

In 2009, Shaun Purcell and a vast team from the Psychiatric Genetics Consortium published a paper in one of the leading science journals,

Nature, arguing that it was not rare mutations but many common variants with small effects that explained the genetic risk for this devastating illness. To support this claim, they developed the first wide-net polygenic index based on 8,008 cases and 19,077 controls. Polygenic indices, such as those created by Purcell and his colleagues, simply sum the GWAS results for all the beads together into a single number for each individual that predicts their likelihood of, in this case, schizophrenia.

Soon the polygenic index approach was employed for a wide range of phenotypes. A *phenotype* is any outcome you can measure—a trait or a disease. PGIs have been developed for phenotypes ranging from height to blood pressure to education. Each one is calculated based on the same DNA loci in our genome. It's just that the value or weight of each locus differs depending on what PGI we are calculating. PGI construction is kind of like a cookbook that tells you how to make one thousand recipes all from the same ingredients by merely adjusting the amount of each one.

Fast forward a decade and a half, and over six thousand GWAS studies have been run on over 3,500 traits or diseases to calculate polygenic indices. As sample sizes increased, predictive power improved. And as that power improved, polygenic prediction soon spread across the field of human genetics like wildfire. In the years since the first polygenic risk score paper was published, over twenty-five thousand have been published using the terms *polygenic risk score (PRS)*, *genome-wide risk score (GRS)*, *polygenic score (PGS)*, or *polygenic index (PGI)*. Over six thousand scientific articles appeared just in the last twelve months. The deployment of this tool shows no signs of abating.

If you can measure a trait in children or adults, you can calculate a PGI for it—anything from cleft palate to sleep chronotype to handedness. So, while the molecular genomics revolution has not yet led to a suite of customized pharmaceuticals to make us taller, leaner, smarter, and healthier, it has led to a new science of prediction. Today, we can predict a US child's (or embryo's) adult height, how far the child will go in school, and whether that child will be overweight as an adult—all from a cheek swab, finger prick, or vial of saliva.

Take education: the first polygenic index trained to predict how far someone went in school was calculated in 2013 based on analysis of 126,559 subjects and only explained 3 percent of the variation. But by 2022, the fourth iteration (EA4 as we called it) could explain 16 percent. As noisy as that is, it's still a powerful tool for prediction: someone in the bottom 10 percent of the education PGI ranking has about an 8 percent likelihood of completing a four-year degree; meanwhile, an individual in the top tenth has about a 70 percent chance of graduating with a bachelor's degree. If you asked me to tell you whether your kid is going to graduate school based on their PGI, I would have a good chance of being wrong. But give me one hundred kids to test and rank order by education PGI, and the odds of predicting graduation outcomes for the top and bottom groups start tilting in my favor based on these stark average differences.

The main way that polygenic indices have been useful to clinicians so far is as a predictive tool—much like family history. Medical professionals are already attempting to triage risk for, say, heart disease, based on polygenic indices. The idea is to prescribe statins—which lower cholesterol levels—much earlier in life to those patients with an extremely high PGI for cardiovascular disease. That's not nothing, in terms of medical utility, but it's a lot less than we had hoped for when we embarked on the Human Genome Project. The lack of an obvious pharmaceutical translation to recent genetic discoveries means that the biggest consequences of the PGI revolution were more likely to be felt in the public health and social scientific landscapes than in medical treatment regimes. Indeed, that is what is happening now.

BUT NOT WITHOUT SERIOUS QUESTIONS ABOUT HOW SUCH WORK WILL be used.

An adage that has been attributed to Irving Kristol, one of the fathers of modern American conservatism, says that a "neoconservative is a liberal who has been mugged" by reality. If that's the case, someone must have really done a job on both Richard Herrnstein and Charles Murray, authors of *The Bell Curve*, published in 1994. Both would have seemed, early in their careers, to be unlikely candidates to author one of the most controversial books of the twentieth century.

Herrnstein, a Harvard psychology professor, has been described as B. F. Skinner's star student. Along with John Watson and Ivan Pavlov, Skinner is considered a founder of behaviorism, a strand of psychology positing that all our actions can be explained by costs and benefits of those actions that we have learned through reinforcement. Put simply, when we get rewarded for doing something, we do it more, and when we are punished, we do it less often. We are totally conditioned by the environment. Skinner went so far as to argue that free will was an illusion. He supported his theories through numerous experiments on lab animals that he conditioned in his eponymous Skinner Box, technically known as an *operant conditioning chamber*. A *Skinner Box* is an environmentally controlled dark chamber meant to house a mouse or other lab animal where the only stimuli the animal receives is controlled by the experimenter. The box contains levers for the animal to press in response to the stimuli. A food dispenser rewards the subject, and an electrified grid in the floor of the box punishes the caged animal.

As a research assistant to Skinner, Herrnstein conducted experiments using the operant conditioning chamber. (He oversaw the pigeons.) His doctoral thesis showed with mathematical elegance how the frequency of actions taken were in direct proportion to the rewards associated with those actions. This steeping in behaviorism wouldn't have seemed to set Herrnstein on a path to coauthor a book that argued for the primacy of genetics in determining who got ahead in American life.

Nonetheless, in 1971, Herrnstein waded into the uproar caused by a 1969 academic journal article on the heritability of IQ. Herrnstein penned a piece in the *Atlantic Monthly* arguing that IQ was largely inherited (biologically)—and thus efforts to improve it or to ameliorate gaps between social groups, were largely for naught.[4] For the rest of the decade, his classes were often interrupted by student protestors. Undeterred, he used the positions he established in his *Atlantic* article to form the core of what would become *The Bell Curve*'s argument.

Unlike Herrnstein, Charles Murray may have always been a conservative deep down. But his early support for labor unionism and the fact that he joined the Peace Corps as a volunteer in 1965 might have led an observer to peg him as liberal. The Peace Corps sent him to Thailand. He stayed there for a few years after his assignment ended and soured on development efforts: he came to see aid programs as counterproductive because they privileged the goals of bureaucrats over those of local citizens and, moreover, because the rapid change brought by such programs undermined local norms and traditions that had evolved over the years. These observations informed his thinking in graduate school at MIT and his 1974 doctoral thesis, "Investment and Tithing in Thai Villages: A Behavioral Study of Rural Modernization."

Murray's breakthrough moment came in 1984 when he published *Losing Ground: American Social Policy 1950–1980*. The book transposed many of the same arguments Murray had applied to Thai villages to the modern-day United States. Murray argued that welfare incentivized long-term poverty. Simply put, if you pay the poor, you induce more people to join their ranks. If you make marriage and job-seeking costly by reducing benefits when people wed or find employment, then you get less marriage and fewer job seekers. Over the long term, social norms—like the expectation to marry or the stigma attached to remaining unemployed for long periods—erode, and a culture of dependency emerges.

Not surprisingly, much debate ensued, and Murray became a superstar on the right. *Losing Ground* shook the foundations of the U.S. social policy debate and directly contributed to welfare reform legislation President Bill Clinton signed a dozen years after its publication. The new law imposed work requirements on recipients that were meant to combat what Murray had highlighted as the perverse incentives of the existing system. Twelve years may seem like a long time, but it is the blink of an eye in the domains of social science and public policy.

It's not hard to see how such an incentive-based account attracted the attention of Herrnstein, the erstwhile operant conditioner. The pair teamed up for five years to write *The Bell Curve*. The 912-page tome has rightly become a byword for bad science and racism. Herrnstein and Murray's central argument was that as a result of Civil Rights legislation and the tearing down of barriers like old boy networks at elite colleges and firms, one's place in society was no longer determined primarily by social background (race and class). Rather, thanks to meritocracy, where we ended up on the socioeconomic ladder was now largely a result of innate ability—that is, our genes.

The authors claimed that inequality was rising not because of tax policies, the decline of labor unions, offshoring, or any of the other developments those on the political left point to. Rather, they claimed that genetic elites were marrying elites to a greater degree than ever before, leading to greater economic disparities as innate advantages were redoubled. Moreover, Hernstein and Murray claimed that society overall was becoming less intelligent since those with lower ability tended to have more children than those with high cognitive ability. Though most of their analysis focused on white people, Herrnstein and Murray argued that the fifteen-point gap in average test scores between Black and white individuals was due to genetic differences and thus not worth trying to narrow through government intervention. This final argument is the main reason for the book's enduring infamy.

Scholars killed forests rebutting *The Bell Curve.* Some of the academic critiques were on the mark; some less so.[5] But the academic community was united in calling the book dangerous pseudoscience. Indeed, Herrnstein and Murray had conducted quite shabby analyses. First, their claims about the increasing salience of genetics (and the corresponding decrease in the importance of social class background) was based on data across a mere seven years of birth cohorts (1957–1964). If a dozen years is a blink of an eye for a social policy idea to jump out of a book and become law, then seven years is a nanosecond in terms of diagnosing a societal trend. It could be those seven years—the tail end of the baby boom—represented a statistical blip. Even if the change in sorting they detected was real, it certainly was not enough to explain the rapid rise in economic inequality that started at the end of the 1960s (and has continued long after *The Bell Curve* was published).

Second, Herrnstein and Murray never actually measured genes at all. Back in the early 1990s, the closest they could come to assessing "innate" ability was to use data from students' performance on a cognitive test *during high school.* Since they measured cognitive ability among teenagers, their results did not show unmitigated genetic potential; rather, these test scores also reflected social (dis)advantages throughout childhood. Third, though at some points in the volume they conceded that heritability is a population-specific concept and contributes nothing to understanding group differences, they ignored this reality and rushed ahead to conclude that race differences in test scores were genetically based.

Regardless of the dubious scientific merits of their claims, *The Bell Curve* spent fifteen weeks on the *New York Times* bestseller list. Its pernicious influence is still evident today. The book's impact can be seen among white supremacist groups that desperately cling to random genetic facts to claim that Europeans are innately superior to other peoples. (One common image in this world is a white person with a milk moustache, since Europeans have the highest rate of lactase persistence—the ability to drink milk into adulthood—a quality that white supremacists argue provides cognitive advantages.) The peal of *The Bell Curve* can also be heard in manifestos like that written by the Buffalo mass shooter in 2022, among other hate crime perpetrators.

Murray has continued to practice bad science. (Herrnstein died in 1994, within days of the book being published.) At first blush, arrival of PGIs to the scientific scene would seem to have solved one of *The Bell Curve*'s major limitations—the lack of any genetic measures. Back in 1994, Herrnstein and Murray *had* to use test scores to make their argument that genes had eclipsed social background in determining who ended up on which rung of the socioeconomic ladder. But three decades later, we have an actual tool for measuring genetics.

It would seem an easy step from calculating a PGI for education or cognitive ability to tabulating the average PGIs for different racial groups and concluding that group differences in IQ are or are not, in fact, innate and intractable. Murray, in his 2020 follow-up to *The Bell Curve*, entitled *Human Diversity*, does back of the envelope calculations to "show" that Black people (who are an admixed population of African and European ancestry, mostly) have lower average education PGI scores than non-Hispanic whites.[6] But this is pure bunk. As in *The Bell Curve*, Murray borrowed a small seed of scientific knowledge and then sowed it in toxic soil, bending and twisting it in unscientific ways to assert a political agenda that appears to have the legitimacy of scientific knowledge but really does not.

The truth is that you cannot use PGIs to make cross-group genetic inferences. That's because genes vary in distinct ways across ancestral populations. For example, the PGI for height—calculated among those of exclusively European descent—predicts Black Americans to be substantially shorter than white Americans, which is patently wrong.[7] In a 2000 article called "Beware the Chopsticks Gene," Dean Hamer and L. Sirota explained part of the problem like this: Imagine that you had a sample of mixed ethnicities. In your sample, Chinese people had a high rate of having Cs at a particular location on chromosome 16—say, 70 percent. But Europeans and Africans in the data only had Cs at a 20 percent rate. This is not uncommon. There are a lot of allele frequency differences across groups due to random fluctuation over the course of many generations. If you ran a GWAS on the outcome, "knows how to use chopsticks," you would find that locus on chromosome 16 would be highly significant in predicting chopstick skills. But it would be a false finding. It's likely that the Cs are not causing anything biological having to do with finger dexterity. They are merely marking an individual as coming from an Asian culture where chopstick use is common. We could actually test this. If we reran the analysis within ancestry groups—that is, conduct a separate GWAS each for Chinese, Africans, and Europeans—we would likely find that the so-called chopsticks gene had no effects within any of the groups. The finding resulted from a cultural difference tagged by a genetic marker correlated with that cultural difference. To avoid this problem, researchers typically run GWAS only on a single ancestral group at a time—for instance, Han Chinese or white British or European. And when you do that, you can't transport the results to another group that lives in a different environment and has a different genetic history.

Scholars continually emphasize that polygenic indices only work within groups, not across groups—that is, the PGIs for height and education developed within white, European samples don't predict well for non-white groups. The result of this valid scientific concern is that over 85 percent of GWAS studies are run on samples with only people of European descent (because there are currently more data on these folks), and only 3 percent are performed on people of predominantly African ancestry. So, the GWAS results we obtain to construct PGIs are based on how those alleles work in white people. Thus, it is scientifically ridiculous and irresponsible to look at PGI differences by racial or ancestral groups because those differences are meaningless. It is easy to dismiss exercises like Murray's on scientific grounds.[8]

STILL, ONE OF THE EFFECTS OF *THE BELL CURVE* IS THE SHADOW IT CAST on all subsequent attempts to use genetics to understand and predict human behavior. The question, then, is why should we pursue this science at all? What benefits will it offer, and will those benefits outweigh the risks it entails—risks of more ideologically driven pseudoscience or willful misinterpretation?

If all genetic analysis can do is predict, rather than tell us about what we can actually do to improve human flourishing, it would seem that all we will get out of the sociogenomics revolution is more bunk like Murray's as well as counterefforts to refute eugenicist pseudoscience. Perhaps we should, as the Russell Sage Foundation did in 2020, pull the plug on funding for this sort of research, at least with respect to outcomes like education, cognitive ability, and other behaviors?[9]

But as I strive to show in this book, there's another use of genetics in studying human behavior. When we see genes and the environment through the lens of sociogenomics—where it's not an either/or but rather a both/and—whole new possibilities to understand the world open up. Take the case of the epidemiology of alcohol use. For the longest time it was thought that moderate alcohol use was healthy. A drink a day (and an apple) keeps the doctor away.

This belief has several origins. Some populations that live a long time—think Italians in Sardinia or Greeks on the island of Ikaria—regularly take a glass of red wine with their meals. Alcohol, and red wine in particular, was so central to the salubrious Mediterranean diet that many researchers, including Harvard longevity scholar David Sinclair, went in search of the magic compound in red wine. However, once subjected to the rigors of randomized controlled trials, the top candidate compound—resveratrol—turned out to be a dud.

If resveratrol wasn't key, perhaps ethyl alcohol itself prolonged life? The most compelling evidence for moderate drinking to be healthy came simply from plotting mortality data by weekly alcohol consumption. Compared to teetotalers, those who consume a drink per day, on average, enjoy 15 percent lower mortality rates. This relationship also holds for specific causes of death, such as stroke. In fact, only when you hit five drinks a day do drinkers exceed the mortality rates of non-drinkers.

This J-shaped graph of mortality by drinks-per-day became the basis for the conclusion that moderate alcohol consumption is healthy. But alarm bells should have been sounding in the heads of epidemiologists. Who doesn't drink? People who are recovering alcoholics, for one. People taking heavy duty medication for cancer or mental illness, for another. In other words, it's not just Mormons and health nuts who don't consume alcohol. It's sick people. If only we had a way to run a randomized controlled trial—the staple of medical and pharmaceutical studies—to assay the true effect of non-drinking.

We can't ethically give some people more alcohol on a random basis. And if we believe that moderate drinking may actually be healthful, then neither can we ethically restrict responsible drinkers from drinking. But what if there was already a natural version of the experiment we wanted to run to know the health effects of alcohol? One running right under our noses. Or, rather, under our skin.

The body, when processing alcohol, first breaks it down, using alcohol dehydrogenase enzymes, into acetaldehyde. Acetaldehyde is toxic and is itself quickly broken down into harmless acetate by acetaldehyde dehydrogenase (ALDH) enzymes. The main ALDH enzyme is encoded by a gene called ALDH2. Some people have a less functional version of this gene, and they take longer to break acetaldehyde down. These folks get what's called the *alcohol flush reaction*, which is very unpleasant. Symptoms include a reddened face (hence the name), nausea, low blood pressure, and migraine headaches. In other words, drinking alcohol with a certain version of ALDH2 is not fun, so people with the less functional version of the gene tend not to drink.

As it turns out, East Asian populations tend to have high frequencies of this version of ALDH2. So, scholar Iona Millwood and her colleagues studied drinking and mortality data from China. First, they found that the same J-shaped curve was present in China as in the U.S.: those who didn't drink at all seemed to have higher mortality rates than those who drank moderately. But they had genetic data, too. They found that, as they suspected, those men with the "bad" gene hardly drank at all. When they compared *these* non-drinkers to drinkers, they found that the former had the lowest mortality rates of all.[10] It was as if mother nature had run an alcohol consumption experiment in a lab. The fact that many non-drinkers were ex-alcoholics or cancer patients didn't come into play here. The only remaining question was whether ALDH2 could be affecting mortality independent of drinking behavior. If, for instance, ALDH2 also raised blood pressure or triglycerides, the lower mortality of those with the nonfunctional version might be due to a lack of those direct effects and not to lower alcohol consumption. Luckily for these researchers, there existed a natural control group: for cultural reasons, women in China hardly drink at all. Chinese women served as their placebo group. Millwood and company compared the mortality rates of women with the good and bad copies of ALDH2 and found there was no difference in mortality for these two groups. They then concluded, rightly in my estimation, that alcohol raises mortality *at any level* of consumption—for everyone, not just for Chinese men.

Their paper appeared in *Lancet*, one of the top medical journals in the world, in 2019. Its findings formed the most important piece of evidence leading to the revision of medical wisdom around alcohol. Today, health guidelines state that alcohol is not good for you, period, and people should, to the extent possible, minimize how much they drink. The public health benefits of this revision will be incalculable.

As much as the paper was revolutionary for clinical thinking around beer, wine, and liquor, it was even more revolutionary for human behavioral science. These researchers used *genes* to study something entirely *environmental*: how an environmental input (alcohol) affects a human outcome (risk of death). If this approach of integrating genes into analyses of human social dynamics could be successful with respect to the question of drinking, perhaps it could yield scientific fruit in any number of domains that have bedeviled social and behavioral scientists: whether going to a fancy college is worth it; how spouses affect each other's moods; how risky behaviors like smoking spread through teenage social networks; and why some of us react very differently to the same challenging situation (say, being drafted into a war) when compared to others in the identical environment. We will explore each of these examples in greater depth in the chapters that follow.

NATURE VERSUS NURTURE HAS BEEN POSSIBLY THE MOST CONTENTIOUS issue in the human sciences for 150 years. On the one side are *blank-slaters*, as those who believe we are entirely shaped by nurture are sometimes derogatorily called by those who think genes matter. On the other side are *hereditarians*, a term sometimes used to disparage people who believe in the deterministic primacy of genes. The fight has been highly politicized and extremely bitter. It has many front lines: Racial disparities. Meritocracy. Fairness. Reproduction. Free will.[11] I have been a participant in this intellectual war for thirty years, first as a "blank-slate" sociology professor studying race, wealth, and inequality; later as a biology PhD, who dove head first into genomic science; and, finally, as a founding member of the field of sociogenomics.

I have emerged chastened, seeing the limits of both social and genetic science but also the power of combining the two fields. This book describes the emergence of the sociogenomic frame, in part by relating my own attempts, over the years, to better understand who flourishes and who flounders in our society.

In casual conversation, I have found that when most people think nature versus nurture, genes versus the environment, they are thinking biology versus society—as I used to. On the biological side are factors such as how quickly our muscles twitch or neurons fire; how strong our bones are; whether we are myopic or hard of hearing; and how elastic the walls of our arteries are. On the social side are questions such as: Were we born into great wealth or penury? How did our mother treat us? Did we have inspiring teachers?

Yet as it turns out, the pathways from a particular sequence of DNA to how well we do on the SATs or what political views we hold or even how much we weigh flow through our own bodies and out into the world. Our bodies, following the instruction of our genes, modify the environment around us, seek out specific inputs, and evoke particular social responses. Genes don't stop at the border of our skin. They affect what kind of environments we experience—that is, how the world treats us. And those environments are critical steps in the long march from DNA to "destiny." With a different environmental topology, the same DNA would lead us to different lands.

At the same time, much of what people have been calling the environment, or nurture, is shaped by the DNA of others. This *social genome* affects us just as other aspects of our environment do, from water quality to the Great Recession. The concept of the social genome should make intuitive sense. The important people in our lives affect us by shaping the nurture we receive. The important people in our lives are partly driven by their own genes. Ergo, the genes of those people affect us in important ways. A straightforward syllogism. But on another level, the social genome represents a profound scrambling of how we think of nature and nurture.

In short, this book will show that much of what we think of as nature works *through* the environment, while much of nurture might be best called "genetics one-degree removed" since the DNA of the important people in our lives constitutes much of our social environment. For so long, geneticists and social scientists have been describing different parts of the same elephant. But by stepping back and taking in the entire beast, we now see human health and behavior in much clearer terms.

Sociogenomics is the application of genetic data to human behavior to gain a more complete picture of nature and nurture. But the sociogenomics revolution goes beyond just weighing the relative inputs of genes and the environment. It has called into question a dichotomy at the heart of the social sciences and human life writ large. Namely, if my genes make their effects clear through the outer, social world while the genes of others influence the social environment I experience, that means that there is no clear line between what we call *genetic* and what we call *environmental*. It turns out that part of nurture is nature, and that part of nature is nurture. The social genome deletes the *versus*.

Sociogenomics can be best described through an image. If you twist one end of a strip of paper and tape it in a contorted position to the other end, you've made a Möbius strip: a looped plane with only one side. If you trace your finger from any starting point on that strip, you will seem to go to the other side of the paper and then back again, without ever having lifted your fingertip from the paper. The social genome is like linking nature and nurture into a single Möbius strip. It's unintuitive and confusing at first, but once we recognize that we live on this Möbius strip of genes and the environment, we can better understand the world around us. By thinking of the interplay of genes and environment in this Möbius-strip way, where genes make up part of the environment, we can paint a much more complete picture of who we are and how we got that way.

I should be clear, however: none of this is to say that by integrating genetics and the social environment, we can now perfectly predict human behavior or that human agency and free will are now things of the past. Rather, sociogenomics gives us a better picture of society and social relationships by recognizing that the boundary between genes and the environment, nature and nurture, sociology and biology, is a false one. What we do with the knowledge produced by sociogenomics—make custom-tailored babies or create a healthier society—is another question altogether.

BACK AT THE FERTILITY CLINIC, I HANDED THE DOCTOR A RECENT ACAdemic journal article explaining the best way to construct a polygenic index. My wife cringed. "We would like to screen based on the genetic risk for ADHD, autism, and schizophrenia," I explained.[12] Besides quelling my anxieties about the risk that I was engendering in our potential child on account of my age, I was also excited about the possibility of conceiving the world's first PGI-optimized baby. I told the doctor that we could write up our case together for a high-profile medical journal—trying to sweeten the pot for her. Alas, at our next visit, she reported back that their labs could not extract an adequate amount of DNA from the embryos for genome-wide testing without damaging them. We would have to fly blind over the genetic terrain. We would implant the winner of the "beauty contest"—as embryologists like to call it—the most symmetrical, nicest looking embryo among the chromosomally normal ones. It all seemed so primitive and unscientific.[13]

As it turned out, the two female embryos we tried did not take. But in 2019, our son was born healthy. (Later that same year, the first PGI-optimized baby, Aurea, was conceived after another fertility clinic pursued what my doctor declined to do.[14]) If not quite created the old-fashioned way (in the back of the proverbial Chevy), my youngest son was neither genetically engineered nor polygenetically selected; his DNA was created from a random, non-chosen amalgam of genes in each of his parents.[15] All the same, the science of sociogenomics will help tell us not only how tall or musical he's likely to be, but also how he will change his parents' political attitudes; how his older siblings may or may not influence his educational path; what kind of spouse he will eventually marry; and how society's reactions to his genetic lottery ticket will close off certain pathways and open others during his long journey along the Möbius strip of nature-nurture.

2

The Racetrack of Life

My father had two passions in life: painting and thoroughbred horse racing. His artwork was a mysterious process to me, even if he frequently dragged me to galleries. His handicapping, on the other hand, we talked about all the time. It was my main means of connecting with him. Technically, handicapping is the practice of adding extra weight to horses that are faster runners, in order to level the playing field. But when horse bettors like my father refer to *handicapping* they mean picking winners based on a systematic approach. He eventually got so good that he qualified for and made a run in the World Series of Handicapping in Las Vegas one year.

Little did I know, but at age six, when I was helping my father plug numbers into a calculator to come up with his picks for the fourth race at Belmot Racetrack, I was being exposed to the nature-nurture debate. When my father annotated the *Racing Form* with his purple and green pens, rating the training each horse received and its breeding, respectively, he was acknowledging the power of both environment and genes to shape winners and losers. In his own way, he was wrestling with one of the most profound questions that has bedeviled humanity over the course of history: Are we who we are because of something encoded within us from birth or because of the circumstances and environments that shape us? My father was working through this tension—between breeding and training; innateness and randomness—every day as he handicapped the *Form*. Moreover, he was competing in a small-scale model that put into relief a deeper question at the heart of our existence, equine *and* human. What exactly makes us triumph or fail?

The matter of nature versus nurture touches other areas of debate. Not only, Is there free will? but also, Is there an essence to who we are, independent of circumstances? Can we rectify the inequalities we see in education, health, or income, or are these differences inevitable? Does it matter how we parent, teach, or coach? Or how we treat criminals? How *much* does our genetic blueprint influence us? *Which* aspects of our environment shape us most?

These questions have dogged us for centuries—since long before anybody knew how genetic inheritance worked or thought much about systemic inequality. Over the years, the pendulum has swung back and forth. In ancient Greece, the sophists thought that people were completely shaped by and developed through education. Character was not something innate but rather learned. Plato, by contrast, believed individuals had attributes and even knowledge they were born with, independent of their environment. In the dialogue "Meno," he posited that education and experience awakened "preloaded" knowledge in a process akin to remembering.[1]

Meanwhile, contrary to what we might assume about early modern racists, many believed that we all possess an underlying potential to be the same. For example, in 1689, John Locke, arguably the most important thinker of the Enlightenment, argued that each human was born as a *tabula rasa*—or blank slate—devoid of any preloaded knowledge or sense of the world. Everything we know, according to Locke, comes from experience, or nurture. This putative sameness of all humans absent environmental differences helped resolve competing ideological strands in Western culture at the time. On the one hand, Judeo-Christian thought viewed all humans as children of God, even if we were born into "original sin." On the other hand, Europeans needed a justification for imperialism. So, when Europeans first encountered non-whites on a large scale in the eighteenth century during the Age of Exploration (and then the Age of Imperialism), intellectuals argued that those who were subjugated in the colonies were inferior in their work ethic or morals or intelligence because their *environments* had failed them, even as all humans remained equal under the eyes of God. In short, blank-slatism allowed for the fundamental ontological equality of all people as well as the differences that were used to rationalize material inequality.

In his 1748 treatise, *The Spirit of the Laws*, French philosopher the Baron of Montesquieu leaned on environmental determinism to explain not just individual variation but the differences between entire cultures and societies.[2] That is, he asserted that climate and geography were the predominant forces in determining human character and societal form. Hot climates, he argued, were conducive to sloth, apathy, and despotism. Heat-induced torpor allowed despots to take advantage of the people, who needed protection from the harsh environment. Colder climes, by contrast, favored industriousness and democracy.

A generation later, the preeminence of the climatic environment in shaping us was extended to race itself. In 1787, Reverend Samuel Stanhope Smith, later the president of what is today Princeton University, published an essay arguing that dark skin was akin to a "universal freckle." Variations in skin tone were like gradations of a suntan, he claimed. In his paradigm, if people from sub-Saharan Africa were relocated to, say, Scandinavia, their skin color would fade over generations to look more like the people indigenous to that region. Ditto, in reverse, for pale-skinned folks who moved to sunnier locales. Such alterations in complexion, furthermore, might also bring about changes in the psychological, cognitive, and social characteristics of these climate migrants. Undergirding this theory was the notion that we all possess an inner potential to be the same if provided identical environments, a rather optimistic theory of human difference.

It took Darwin's theory of evolution by natural selection to disrupt this intellectual equilibrium and push the pendulum away from its nurture apogee and back toward the importance of nature—of innate characteristics fixed at birth. Darwin contended that acquired (i.e., shaped by nurture) traits could not be inherited; instead, he asserted, change only occurred through descent with modification and natural selection.

He observed how farmers bred livestock and plants to achieve specific ends and theorized that mother nature did this as well. Most plants and animals had many offspring (his principle of overproduction). These offspring were different (variation). Not all survived and reproduced. Only those most adapted to the environment would make it. These variations, in turn, were heritable; that is, they would be passed on to descendants of the survivors. The word *mutation* was not added to the biological lexicon until 1901. And the word *gene*, the idea that there were discrete physical packets of important information that controlled traits and were inherited, didn't make its appearance until eight years after that. But already in the late nineteenth century, Darwin recognized that biological change in traits was much more gradual than had been assumed by Stanhope Smith and his ilk, who implicitly subscribed to a theory of inheritance called Lamarckism. Jean-Baptiste Lamarck had argued in 1809 that acquired traits could be passed on to offspring. That is, if a horse stretched its cervical vertebrae over years of browsing the highest leaves on a tree, it would pass on a longer neck to its descendants. If a human moved to a climate with more sun, they would pass on their tan to their children.

Darwin rejected the inheritance of acquired traits; his account of evolutionary change took place over a much longer period than someone like Stanhope Smith had envisioned with his metaphor of the universal freckle. In fact, Darwin's scientific revolution opened the door for the idea that humans were *not* all the same once you wiped away the effects of climate and other environmental influences.

Long before Darwin's theory, racial pseudoscience had already been seeking biological bases for classifying humans into distinct groups. In 1795, Johann Friedrich Blumenbach, considered the father of anthropology, divided humans into five races based on analysis of his large collection of skulls. (It is from Blumenbach that we get the term *Caucasian*, his other four races being Mongolian, Malayan, Ethiopian, and American.) The scientific community soon found itself embroiled in a heated discourse: monogenism versus polygenism. At the heart of this debate lay the question of human origins: Were Blumenbach's races all part of one unified species, or did we arise from separate evolutionary lines? Monogenists, including religious traditionalists who clung to the notion that we were all God's children, argued that we were one. Conversely, polygenists asserted that different races were, in fact, distinct species—notwithstanding the empirical evidence that interbreeding by race begat fertile offspring, a common definition of species boundaries in sexually reproducing organisms.[3] In an interesting irony, Darwin aligned himself with the religious conservatives, who were otherwise at odds with him, arguing that all evidence pointed to the fact that all humans formed part of the same species. When genetics comes into the picture, it can often scramble ideologies and political allegiances.

The monogenists eventually emerged victorious in this debate. However, even though the scientific and public consensus was that we were all one species, Darwin's notion that the physical traits we observed among people living in different parts of the world took thousands or tens of thousands of years to emerge had a lasting impact on how we viewed human difference. The environment took a back seat—after all, the universal freckle theory had been shot down—and innate biological differences were now seen as the driver of not just physical differences but of behavioral and social differences as well. In other words, if evolution over generations explained why some people had broad noses and others narrow ones or why some people had long limbs and others short, then it might explain why some groups were rich and others were poor, why some were colonizers and others were imperial subjects, why some cultures were literate and others not.

It's important to note, however, that in Darwin's paradigm—and indeed in modern evolutionary theory—there is no direction to evolution. There is no species that is "higher" or "better" than another. We may like the descriptor, "the lowly earthworm," but other than living on the ground, there is nothing biologically about the worm that makes it "beneath" us. Each species is optimized to its environment. Humans would epically fail if our task was to create topsoil and live off the droppings on the forest floor, just as earthworms would be terrible at milking goats. There is no such thing as a *natural* hierarchy. Hierarchies are social constructions imposed on the tree of life—or, on the human family tree. That observation, however, flew in the face of Judeo-Christian thought that put humans above the rest of God's creatures. Moreover, it didn't jibe well with European supremacists who wanted to insist on a hierarchy of human races, with white Europeans on top.

Darwin's assertion of a nonhierarchical nature to evolution didn't stop a rash of thinkers from selectively (and wrongly) applying evolutionary theory to explain human inequalities. Just as the earthworm was adapted to life in the dirt, white supremacists argued that Europeans, Asians, Native Americans, and Africans, were all *biologically* adapted to their surroundings thanks to evolution. Europeans, they claimed, were biologically suited to managing the rest of the world, to a life of literacy and modern conveniences, while Africans were suited to hard physical labor. But these claims flew in the face of evidence that, for instance, literate cultures independently evolved in various parts of the world. Or that non-European languages were just as complex as European ones. Or that plenty of non-Europeans functioned very well in European society when they immigrated.

The misapplication of evolutionary theory to explain human differences was forcefully advanced by Herbert Spencer. Spencer, the child of a religious dissident, was a radically progressive thinker in his time who advocated for female (and child) suffrage in his native England and who sought universal scientific laws to explain all of human life in defiance of ecclesiastical accounts of existence. (He also invented the precursor to the paperclip.) He is widely considered the leading intellectual of the last decades of the nineteenth century. It is not surprising, given when Spencer wrote, that Darwin's theories formed a centerpiece of his effort to explain human societies. It was Spencer, not Darwin as is commonly believed, who coined the phrase, "survival of the fittest" after having read Darwin's *On the Origin of Species.* In his 1864 book, *Principles of Biology,* Spencer applied the concept of evolutionary competition to how societies functioned and who got ahead in them—the proverbial winners and losers.

Spencer was not alone in taking the Darwinian ball and running off the field with it. Darwin's own half-cousin, Francis Galton, one of the founders of modern statistics, was the first to analyze data on humans within the framework of Darwin's theories in his 1869 book, *Hereditary Genius.* Galton was even more of a polymath than Spencer. Not only was he the founder of modern statistics and psychometrics (the science of measurement of psychological traits), but he was also a renowned geographer and published the first-ever weather map. To top that off, he is the father of fingerprinting—both the system of fingerprint classification and the use of fingerprints as a unique form of identification, now commonly used in criminal justice systems. Galton even gave us the phrase "nature versus nurture," which set the terms of debate for human science for the next 150 years. Galton himself fell squarely on the side of "nature." In *Hereditary Genius*, Galton showed that mental traits varied in much the same way as physical traits (like height) and that they tended to cluster in families. Just as tall parents were likely to have tall offspring, Galton showed, criminal parents were more likely to bear criminal offspring and highly intelligent parents were likely to produce smart kids.

Galton's fundamental flaw was to attribute the clustering of outcomes within families to genetic inheritance while ignoring the environmental differences between households. That is, when tall parents produce tall children, it could be because rich families have better access to nutrition in both the parental and the offspring generations. Likewise, poverty and a lack of opportunity for both parents and children may cause both generations to veer into a life of crime. While we now know that variation in height is mostly genetic, when we consider the other, more controversial outcomes that Galton examined—intelligence and criminality—environment may play a much greater role.

However wrong his models of human difference were, Galton saw that Darwin's theories discredited many premises of Christian theology—that all humans were a fixed deal, conceived in heaven as the children of angels (even if the Bible itself distinguished between races). If we were not born from the mind of God but instead had evolved over millennia here on Earth, shaped by the forces of natural selection, why couldn't we, ourselves, have a hand in that shaping? Could we do God's work, so to speak, Galton asked, bettering the human condition through selective breeding of humans in the same way crops and domesticated animals had been cultivated since the Neolithic Revolution? This would occur in a process he called *eugenics* in his 1883 book, *Inquiries into Human Fertility and Its Development*.

During the Neolithic Revolution, approximately ten to twelve thousand years ago, some humans radically departed from their hunter-gatherer way of life. They settled down from their nomadic lifestyle, domesticating plants and animals. This meant that they collected seeds and planted them; when their crops grew, they selected the seeds from the best yielding plants and planted only those seeds the following season. A similar process happened with the breeding of goats and sheep. Controlling which plants and animals reproduced gave humans the power to shape the genomes and the population characteristics of their food sources. It was an enticing logic: Why not apply the same approach to us?

Galton was mostly interested in "positive" eugenics—that is, encouraging successful people to have more children—rather than "negative" eugenics—preventing people deemed unfit from reproducing. However, from the get-go, Galton's notions were infused with racial hierarchy: "The most merciful form of what I ventured to call 'eugenics' would consist in watching for the indications of superior strains or races, and in so favouring them that their progeny shall outnumber and gradually replace that of the old one."[4]

In 1913, the United Kingdom passed the Mental Deficiency Act (which superseded the Idiots Act of 1886). This law sought to institutionalize those with mental deficits, segregating them from the general population. It did not, however, provide for sterilization. A bill that proposed sterilization was brought before parliament in 1931; it was resoundingly defeated, 167 to 89. Again, the politics of genetics was complicated: opposition to the bill came mostly from Labour MPs and from Roman Catholic MPs, who opposed any form of birth control.

As a result, the harshest, most coercive form of eugenics was not taken up in Galton's native UK; instead, it came to most prominence in the U.S. during the early twentieth century as immigrants were pouring into American cities at an unprecedented rate. Charles Davenport was a Harvard biologist who led the U.S. eugenics movement and focused, unlike Galton, on negative eugenics—the discouragement or outright prevention of people *he* deemed unfit from reproducing. And while the British eugenicists thought that interracial mating might have benefits, Davenport coauthored a pseudoscientific book arguing against misce-genation, entitled *Race Crossing in Jamaica*, which was so bad that it was panned even by one of Galton's successors in the eugenics movement, Karl Pearson. Davenport and the other U.S. eugenicists had a big influ-ence on public policy. Their arguments contributed to the passage of the restrictive Immigration Act of 1924, which established national quotas favoring Northern and Western European countries of origin. They also influenced the passage of thirty-two state laws authorizing the government to sterilize people deemed "unfit" between 1907 and 1937; an estimated total of sixty-four thousand people were forcibly sterilized as a result of the passage of these laws.

It might not be surprising that an ideology (and practice) like eugenics gained traction when one population (native-born citizens) perceived a threat from another (immigrants). What might be harder to believe is that eugenics formed a central plank of the Progressive Movement in the early 1900s. It's important to keep in mind the ethos of the time. Great scientific progress was being made in transporta-tion, factory production, electricity and energy, and medical science. The application of science to agriculture—most notably with the application of synthetic fertilizers to crops—was leading to increased food yields. Scientific breeding had also become key to improving food supplies. It perhaps seemed a logical next step to apply the same principles of science to better the human species—whatever "better" meant—especially when contrasted to religious doctrines that dis-couraged contraception and kept women in a state of maximal fertility.

Indeed, feminists such as Margaret Sanger—famous for cham-pioning women's access to birth control—got on the eugenics band-wagon. Sanger had been raised as one of eleven children born to her overworked mother. This childhood background, in addition to her

experience as a nurse who regularly encountered the health consequences for women of unfettered fertility, led her to her lifelong
crusade for access to contraception. But she went farther than just
advocating for this access. Sanger thought that "feeble-mindedness"
(i.e., intellectual disability) was associated with "abnormally high
rate[s] of fertility," which led to a "biological menace."[5] She argued
for the segregation of intellectually disabled women and the sterilization of their male counterparts to make sure that "parenthood is
absolutely prohibited to the feeble-minded."[6]

During the same period Sanger was advocating for these policies,
in *Buck v. Bell*, the Supreme Court upheld the notion that states could
sterilize people deemed "unfit" for the "protection of the health of the
state." Otherwise considered a liberal intellectual giant in Supreme
Court history, Justice Oliver Wendell Holmes Jr. wrote the majority
opinion allowing for the sterilization of Virginia woman Carrie Buck.

In this way, eugenics was thought to be an effective tool to battle
the Malthusian trap, in which fertility became so high that everyone was on the verge of starvation. (Thomas Malthus was the English father of demography who argued that whenever agricultural
advances led to improved food supplies, people responded by having
more babies or, at least, more children who survived; such rises in
population inevitably brought the per capita food supply back down to
its original, barely subsistent level.) The most forgiving and idealistic
interpretation of Sanger's and the Progressive Movement's embrace
of eugenics was that controlling who had babies was just another tool
in the social welfare toolbox that also included expanded schooling,
programs to assimilate immigrants, civil rights, election reform, and
the prohibition of alcohol—to name just a few. However, in practice, defining who was "fit" to procreate and who was "unfit" much
more reflected biases about race and class than any objective scientific
assessment. (It perhaps was not a coincidence that Sanger's contraception clinics were disproportionately sited in Black communities.)

Besides the same misinterpretation of family resemblance that Galton had committed, the main empirical "evidence" that advocates of eugenics advanced during the Progressive Era was that immigrants had lower test scores than native-born U.S. citizens. At the time, "professional researchers recorded that '79% of the Italians, 80% of the Hungarians, 83% of the Jews, and 87% of the Russians are feeble-minded.'"[7] This, of course, ignored the fact that immigrants at the time happened to come from Eastern and Southern Europe and didn't speak English as well as native-born U.S. citizens or earlier waves of immigrants from English-speaking countries, hence their lower test scores. This is the core mistake that hard-core hereditarians and eugenics-types make over and over: they assume that you can measure someone's inherent "natural" capacities without capturing the discriminatory social environment that may be impeding those same capacities.

Making this same error, more and more politicians got on the eugenics bandwagon at the start of the twentieth century. The first sterilization law in the U.S. was passed in 1907 in Indiana; the most recent (and hopefully the last) was passed in Georgia in 1937. Most have now been repealed. While programs of *mass* sterilization tapered out in the United States, the Nazi party in Germany picked up the baton, explicitly looking to programs in the U.S. that sterilized criminals and those otherwise deemed unable to care for children as a model for their agenda of eugenics and racial cleansing. To this end, the Nazi regime cribbed much of their 1933 Law for the Prevention of Hereditarily Defective Offspring from U.S. state laws. As a result of this German statute, thousands of people were sterilized leading up to and during World War II. The Nazis went further, of course, "euthanizing" many others—even before their genocide of European Jews and Roma.

The pendulum continued to swing back and forth. Largely in response to the World War II atrocities committed in the name of

ethnic purity, by the second half of the twentieth century, the pre-
vailing view among elites was that all humans were, at their core, the
same—our differences were merely cosmetic. This view was perhaps
best represented by a 1955 exhibition at the Museum of Modern Art
titled "The Family of Man." The exhibit consisted of 503 photographs
by 273 artists from 68 countries, depicting humans from all corners
of the world, grouped thematically around topics like love, death,
children and so on. The Pulitzer Prize–winning poet Carl Sandburg,
who wrote the text that accompanied the images, encapsulated the
era's ethos in his introduction:

> There is only one man in the world and his name is All Men.
> There is only one woman in the world and her name is All
> Women. There is only one child in the world and the child's
> name is All Children.

The exhibition traveled to over 150 countries and was reproduced in
a book by the same title that has sold more than four million copies
and has never gone out of print. UNESCO has even bestowed world
heritage status on the collection. The fundamental message: we are
all the same, at least when it comes down to what matters.

Around the same time "The Family of Man" was touring the
world, horse racing was undergoing its own pendulum swing toward
blank-slatism. Kelso was a male horse with a relatively undistinguished pedigree.[8] As a foal, Kelso was scrawny, and he was so hard
to handle that he was gelded in the hopes of calming him down. He
raced only three times before age three. Despite his relatively unex-
ceptional genetic profile and seeming lack of potential, under the care
of trainer Carl Hanford, Kelso blossomed in his third year, winning
nine of ten races that year. The following year, he won seven of nine.
He tied Man o'War's record for the mile-and-five-eighths distance.
And he still holds the world's record for a two-mile-long dirt race. He

was named horse of the year five times—his nearest competitor held the title three years—and his almost two million dollars in earnings at his retirement in 1966 due to a hairline fracture in his hind foot represented the highest amount for a horse in history. (The Triple Crown winner Affirmed surpassed Kelso in 1979.)

At the time, Kelso's humble beginnings were taken as evidence that nurture trumped nature in thoroughbred racing and that horse players should pay a lot more attention to the trainer than to a horse's breeding. This aligned with the general view on humans: genetics was responsible for differences that were largely superficial and had little to do with an individual's success. As it turned out, the story was a bit more complex. Today, the best estimates are that genetics explains about 40 percent of overall thoroughbred performance; that is, differences among genes in racehorses explain about four-tenths the variation in times.[9] The effect of genes seems to be the greatest for middle distances—and declines precipitously for those long distances at which Kelso excelled.

When my father was handicapping in the 1970s and 1980s, "all nurture" was still the dominant paradigm, giving my father, who painstakingly checked the genealogy of horses in the guide he had purchased by mail order, a perceptible edge. My father even discounted the power of trainers, suggesting that some trainers were made to look better than they were because they got the best-bred horses on account of their fame. This observation distilled, in horses, the fundamental problem of "self-selection" I would struggle with for most of my career studying humans. Namely, if the best trainers got the best raw material to work with, how could we ever know the true value added of their nurture?

This nature-versus-nurture question is so complex that it might seem intractable. But people like my father believed there must be a way to master it through cold-eyed statistical analysis. With enough color-coded pens and solar-powered calculators, we might find an answer—at least an answer for the seventh race at Aqueduct on a given Spring afternoon.

MY EXPOSURE TO HOW NATURE AND NURTURE SHAPE OUTCOMES WAS not limited to horses. The Lower East Side, where I grew up, was a poor neighborhood. My parents were bohemians who relied on food stamps in their quest to make it as a painter and a writer, respectively. We lived on the twenty-first floor of a low-income housing complex that was named—for reasons unknown to me—after Tomáš Garrigue Masaryk, the first president of Czechoslovakia. The yellow-brick structures housed more than five thousand souls within earshot of the J trains that rumbled over the Williamsburg Bridge. Graffiti covered every surface, brown and green shards of malt liquor bottles littered the playgrounds, and garbage bags spilled out on the curbs. Crime was rampant. At one point, my parents chained the TV to the radiator to prevent its theft. My karate teacher was shot in the head and killed when I was eight. Our steel front door felt like the reinforced gates of a medieval fortress.

At the same time, I had a marvelous childhood of street games like ring-o-levio, caps, manhunt, stoop, and stickball that are all just about extinct in today's world of apps and screens. My neighborhood had five main groups—Puerto Ricans, Dominicans, African Americans, Chinese, and a few remaining Jews (they had once been the dominant ethnic group in the area)—and our neighbors were warm and open, a far cry from the stiff niceties of suburbia. They were also mostly generous, despite being poor. Block parties were not uncommon, and everyone looked after one another's kids in the playground. There was a babysitting club where, my mother claimed, I cost double chits because I was such a handful. Once, my sister got her fingers crushed in the front door of the building, and a resident picked her up and raced with her in his arms—my mother trying to keep up with her pigtails flapping as she ran—to the nearest emergency room, which was actually pretty far away.

Though we shared our neighbors' day-to-day material conditions, we were middle class. That is, while I may have received free public-school lunch, I had a degree of "cultural capital" that paved the way for me to eventually escape the neighborhood, gain access to higher education, and realize my professional dreams. I had a mother with a college degree, and other relatives with doctorates. Doctors, professors, and other professionals regularly came to our apartment for dinner and engaged me in conversation about what I was learning in school. I also had exposure to a wider world: every summer, when things got hot and tense in the neighborhood, we decamped to Susquehanna County in rural Pennsylvania, where my mother was from and where my grandparents still lived. Not to mention the fact that the color of my skin afforded me a degree of protective privilege that most of the other kids on Avenue D did not enjoy. In fact, I was the only student my first-grade teacher did not beat—on account of being white.

I was acutely aware of economic inequality and keenly interested in escaping the poverty of my neighborhood. I started working odd jobs at age nine and investing that money in the stock market by age eleven—by which time I had programmed a stock-picking system into the floor-model computers that the local Radio Shack franchise allowed me to use gratis after school. When I was twelve, I snipped alligator logos off my grandparents' golfing shirts and sewed them onto my own generic shirts to pass them off as Izod Lacoste polo shirts. Aware of the vast differences in income and wealth that I saw between families, I wanted to ensure that I would have a level of economic security that my parents had mostly forsaken with their career choices—even if at the time I would not have been able to articulate this motivation in those terms.

As I got older, I seemed to be on track to economic security as my cultural advantages manifested. For instance, when my local school went from bad to worse, an artist friend of my parents allowed us to use his address in tony Greenwich Village to lie to the Board of Education about where we lived. All of a sudden, in third grade, I was attending school on the rich side of town, where kids showed off by quizzing one another on the meaning of "antidisestablishmentarianism" and where some families owned entire brownstone buildings with actual backyards that weren't riddled with broken glass and discarded tires.

I never attended school on the Lower East Side again after that. By the time I started high school, my parents' names had come up on the waiting list for subsidized artists' housing—in the very same building we had used as a fake address a few years earlier. And thanks to my performance on a single test, I managed to squeak into Stuyvesant High School, the magnet school that is still seen as the gem of the New York City public school system. From there I left New York City for a while, attending the University of California, Berkeley, before returning to complete graduate school at Columbia University. My jobs since then have been at elite institutions. As a tenured professor, I enjoy the ultimate level of job security along with an unmatched level of autonomy. When one of my kids gets sick, I can reschedule my classes and meetings for another day. My family has always had health insurance. I own my home. I can afford healthy, organic foods, and I can show my children the world.

Today, most of the kids I knew from the old block work in low-wage service sector jobs. They work for slightly over minimum wage and enjoy little to no control over their work schedules—they may find out they have to work the late shift the day after tomorrow or that they aren't needed at all next week. If they provide health insurance at all, their employers offer high-deductible plans where almost all the cost is born by the employee. One kid, whose father coached our neighborhood's Little League team, *Los Piratas*, ended up doing almost two decades behind bars on a drug charge. Another kid I knew, who was chosen to ride on Big Bird's back for the opening credits of Sesame Street, was killed in a car crash when the steering wheel crushed his chest because the vehicle had no air bags.

I don't need a fancy scientific study to tell you that my family's race and class background played a big role in why I am teaching at an Ivy League university while most of my childhood neighbors are working low-wage service jobs. When you pull back and look at America's population as a whole, you can see the invisible contours of race and class clearly etched into the social landscape, guiding us to different outcomes. One kid getting arrested or a karate teacher being shot is a tragedy, but when these outcomes happen over and over to a group of people, that's likely evidence of the pernicious effects of poverty or racism in America. When we take the thirty-thousand-foot view, we see that the paths we walk in life are, in fact, well-trodden grooves. *On average*, it's no surprise that a white kid whose mother graduated college made it out.

But there were also outcomes among my neighbors that *were* surprising and which couldn't be easily predicted from a statistical model involving only race and class. Leila, who lived a couple buildings down from me, had, on paper, many of the same advantages I did. She was white, and her mom had attended college. But the psychological effects of violence and other social ills of our neighborhood were far worse for her than for me. By seventh grade, she was skipping school every other day, and by high school, she had developed a fear of leaving safe environments. She couldn't leave her family's apartment. This agoraphobia was, to some degree, a rational response in an environment where sometimes people were struck by bottles thrown out of high windows, but as a result, Leila didn't complete high school. For a while she worked as a taxi dispatcher—a job she could perform with a phone line in her bedroom—but eventually that work dried up. When her mother decamped for the warmer weather of the South,

Leila went with her. Today she takes care of her octogenarian mom in Austin.

Leila's younger sister Crystal, on the other hand, did finish high school.* Despite growing up in the same adverse conditions, she went on to college at the University of Chicago, and then to graduate school in animal science. Today she works high up in the research arm of the USDA. Even with the same parents, the same household environment, and the same external conditions, Leila and Crystal, as individuals, refracted those conditions quite differently. In other words, the question of why some kids got out of the neighborhood and others did not cannot be answered exclusively by the obvious forces of race and class. Sociological categories are critical for understanding group differences—like the fact that women are paid 78 cents on the dollar to men or why Black men earn 87 cents to the white male dollar. Structural disadvantage, racism, sexism, and other social forces are at play here. However, by definition, these categories don't tell us about within-group differences, even when we take an intersectional approach and talk about subcategories like gay, Latino men. That's where we need another tool.

Along the same lines, sometimes outcomes among my neighbors were even more surprising because they flew directly in the face of these group, structural forces. My good friend Jerome McGill, or Jerry, lived in the public housing projects a couple blocks away with his mother and sister; in middle school, we rode the M14 bus across town together. Jerry was outgoing, always smiling, and ever polite to adults. He could have been the junior mayor of our neighborhood, given how universally liked he was. We mostly played Asteroids, whiffle ball, or caps together. But on New Year's Day 1982, he was hanging out with a friend on the stoop of a tenement down the block from his building when a shot rang out from a crack house across

---

* Leila and Crystal are pseudonyms; all other names in this manuscript are real.

the street. It ricocheted off a wall and lodged in his neck, injuring his spinal cord and confining him to a wheelchair. Despite the challenges of his life, he is now a successful author and disability activist. Perhaps, I'd venture to speculate, his indomitable spirit was encoded in his DNA and kept him from succumbing to the tragedies that littered his life.

Marc Ramirez—the kid whose dad coached *Los Piratas*—served more than eighteen years of his twenty-year prison sentence, with time off for good behavior. Over those long years, he did not give up on making a future for himself. He managed to maintain a relationship with his two sons. He studied law while in prison, even arguing a case on prisoners' rights that went all the way to the U.S. Court of Appeals. (Had he not been released—and thus lost the standing to refile the case—Marc could have taken the case to the Supreme Court.) Once out, Marc enrolled in law school and worked in the Bronx public defender's office; after graduating, he successfully sought a way to be admitted to the bar, despite the felony on his record. His story is a reminder that our circumstances do not determine what we can achieve. For most people, prison is an irredeemable trauma. Marc paid the harshest possible consequences for his youthful mistake. But he lives his life with hope and kindness, working to prevent what befell him from happening to others.

Are Leila, Crystal, Jerry, and Marc such "exceptional" cases? Wherever we look, we can see people who react very differently to similar environmental circumstances. But why is this? To return to the case of Leila and Crystal: How do we explain why one sister managed to thrive despite her circumstances, while the other succumbed to them? It is theoretically possible that the reasons have everything to do with nurture. Perhaps their birth order or age difference resulted in different perceptions; or perhaps their mom treated them differently based on things going on in her own life when they each reached adolescence; or it could have been that Crystal happened to have had a magical science teacher in eighth grade who productively channeled her anxieties about urban life.

But there is also nature. Today we know that two-thirds of the variation in agoraphobic tendencies tracks genes—which would seem to point to a biological explanation: the reason for their polarized reactions may lie in their respective DNA; full siblings share half their genes on average, but for any given pair, genetic similarity can run anywhere from 35 to 65 percent.[10] Our social circumstances alone can't explain why often people from the same family, the same neighborhood, the same school, in the same era turn out vastly different. Why do some of us succeed? Why do some of us flounder? The answer is nature *and* nurture, genes *and* the environment. Genes affect an individual's chosen environment, and, in turn, environments affect which genes are most important in life. Only by considering both can we begin to understand what makes us who we become.

TODAY, I TEACH AT PRINCETON UNIVERSITY, THOUGH I STILL LIVE IN New York City, not that far from my rapidly gentrifying childhood neighborhood. I have now spent a good thirty years (and counting) trying to unpack the puzzle of my childhood and those of my neighbors, scientifically and statistically—namely, I have been trying to answer the question of how socioeconomic status is transmitted across generations. Put another way: Who thrives and why? I am still working each day, with more and more data, to better understand the interplay between nature and nurture that makes us who we are.

Only recently did it dawn on me that this interest in picking winners, so to speak, was a direct bequest of my father's. What he did for *Equus caballus* with colored pens, spiral notebooks, and the *Daily Racing Form*, I was pursuing for *Homo sapiens* using microcomputers and large-scale datasets. We were both trying to unpack the forces that led a particular organism to a given outcome—success in a horse race in his case, success in society in mine. Moreover, he and I were both trying to find the particular creatures that beat the odds, that performed better than might have been predicted by the more casual observer.

We generally don't like to think of human life as having winners and losers in the same way that six-furlong races do. It feels reductive, or crassly capitalistic. It's certainly true that we all have different goals and values; unlike eight thundering horses, we are not all aimed at the same finish line. And yet most of us, in the aggregate, do yearn for the same things: We want ourselves (and our children) to be healthy, to lead fulfilling lives, to have meaningful relationships, and to be relatively free from the stress of economic insecurity. We want to flourish. The trouble, of course, is that not all of us do, and it's impossible to ignore the fact that how much education we get, how much money we have, how much power we hold—just like pedigree and training—all strongly influence the degree to which we prosper.

We move in a world that is more like a horse race than we might care to acknowledge. But if we can better understand the variables, perhaps we can also better manipulate the odds so that more of us are able to succeed, much like horse race officials try to level the playing field with different weights for each horse. We should care about the invisible social and molecular forces that influence who thrives and who struggles because with this knowledge, we can fine-tune our policies to improve the life chances of each person and improve our collective outcomes. In short, with a better understanding of how nature and nurture operate as one, we can make sure more people fall into the thriving category. Maybe, with this knowledge, Leila could have received the kind of support for her agoraphobia—psychological, medical—that would have ensured she could expand her job prospects and start a family. Perhaps, with this knowledge, there could have been sounder public policies to help the heroin junkies of my neighborhood beat their addictions. Knowing what causes people to end up where they do is the first step to changing where they land, or to making those landing spots better places.

In retrospect, I find it strange that my father and I never discussed the commonality in our pursuits. By the end of high school, I had programmed his entire handicapping system into a BASIC computer program that I cobbled together on my Commodore 64 and stored on a cassette tape. I taught him how to plug in the necessary data from the *Form* to compute the predicted odds for each horse. And then I boarded a plane for college in California, thinking I was leaving the art and science of picking winners back in New York. By the time I became aware of the parallels between our interests, my father's mind had eroded due to Alzheimer's disease.

Had my father seen the connection between our callings? And if he did recognize the similarities between human science and horse handicapping, had he pondered how the interaction of breeding and training applied to his own life? His mother, a curious contrarian, had taken up drinking when Prohibition was instituted because she was intrigued by what the fuss was all about. Did he wonder whether her drinking and smoking during pregnancy affected his health and career success in the world? Or whether his economic struggles were baked into his genes? His brother, six years younger, ended up much like me: an Ivy League professor. But his initial motivation to go to graduate school was to avoid the Vietnam war, so could his career really be attributed to a difference in nature?

Perhaps my father's chances of successfully handicapping horse races will turn out to be greater than our success in predicting human flourishing. After all, today, only white supremacists compare humans to thoroughbreds in what's called "racehorse theory"—an argument for selective breeding (i.e., eugenics). "You have good genes, you know that, right?" Donald Trump told a mostly white crowd of supporters in Bemidji, Minnesota, during the closing weeks of the 2020 Presidential campaign. "You have good genes. A lot of it is about the genes, isn't it? Don't you believe? The racehorse theory. You think we're so different? You have good genes in Minnesota."[11] Ten years earlier, he told CNN, "I'm a gene believer. Hey, when you connect two racehorses, you usually end up with a fast horse."[12]

Notwithstanding Trump's confidence, perhaps it's too hard to handicap human beings in a race with no clear starting gate and no agreed-upon finish line. That is, maybe the controlled world of thoroughbreds is a more satisfying pursuit to devote one's life to. If my father were still around, though, I would argue with him that the very chaotic nature of human striving makes it, ultimately, a more interesting game to calculate the odds on.

3

What's Your PGI?

After finishing that BASIC computer program that implemented my father's algorithm, I thought I was done handicapping, but I couldn't seem to quit. I left for Berkeley in August 1986, excited to reinvent myself in the sun-drenched Golden State. But in those early semesters, I floundered a bit, trying on all sorts of identities. I bought a motorcycle. Joined ROTC. Boxed for the school team. In my unsettledness, I found solace in weekly trips to the Golden Gate Fields racetrack. My dad's system didn't seem to work as well out West—maybe because the air was drier, I don't know—but I did make enough in winnings that I fantasized about a career after graduation as an itinerant horse handicapper, hopping freight trains across the West to alight at various thoroughbred racetracks.

Though I couldn't have expressed it at the time, my restlessness was animated in part by a desire to understand where I'd come from and in what ways this mattered. I took pre-med classes, because being a doctor was about helping people, but I also took history and philosophy classes, because helping people required understanding them. In the end, I cobbled together my own major—Art and Technology in the Twentieth Century—and charted how the sweep of history, and all its coincidences and randomness, had shaped our shared culture and, in turn, individual lives. Believe it or not, this involved researching how Looney Tunes reflected the evolution of American self-image: from the Depression-era insecurity of Porky Pig to the militant Daffy Duck of World War II to the confident Bugs Bunny of the post-war Pax Americana.

On my trips to the racetrack, I read Charles Bukowski, William Burroughs, and other writers who glamorized the lifestyles of the down and out as the path to cogent social observation. I wanted to write gritty novels like they did, but after I graduated, I landed contract work as a stringer in the aftermath of the Gulf War in 1991. I found myself in Egypt reporting on Arab feminism, civilian prisoners detained in Iraq, and environmental degradation of the Nile by cement factories. I loved the stimulation of living in a different culture, but journalism did not afford me the opportunity to ask deeper questions about how and why what I was seeing around me—the Lebanese Civil War, the division of Cyprus, Assad's iron grip on Syria—had come about. I liked the analysis I had done for my senior thesis: not just describing the social world around me but coming up with a theory of why it was so. Perhaps the right theory could help alleviate the poverty I'd seen, whether in the slums of Cairo or my childhood neighborhood.

This is how I ended up back in New York, in a program in public policy. But there I found that I was being trained not to ask the *how* and *why* questions but rather to make wise administrative decisions. We spent a lot of time on cost-benefit analyses. We learned about total quality management (TQM), an approach from Japan that was all the rage at the time. We read many case studies. (One I remember vividly was about a roach elimination program in public housing.) These were all worthy pursuits, but they did not satisfy my desire to understand how social problems developed in the first place. I wound up pivoting into a PhD program in sociology.

And there, at last, I found what I'd been looking for. My college fiction writing professor, Leonard Michaels, had said that the difference between a journalist and a fiction writer is that upon coming to the scene of a car accident, the journalist jots down the basic facts—how many cars, how many injured? By contrast, surveying the same scene, the fiction writer is drawn to a particular gnarl in the tree that one of the autos wrapped itself around. That gnarl is the most relevant detail through which an account of the event emerges. Meanwhile, the sociologist, I learned—in contrast to both the fiction writer and the journalist—focuses on the public policies that led to the proliferation of SUVs on the roads; the class, gender, and race of the drivers; and the fact that, the grisliness of the scene notwithstanding, the truth is that motor vehicle accident rates have been falling steadily for decades—though slower for the U.S. than other rich nations as of late. Sociology, in the words of C. Wright Mills, a towering twentieth-century figure of the field, "connects personal problems to public issues."[1] Put another way, it reveals hidden social structures that affect our individual lives.

Sociology gave me rigorous methods to investigate important social questions. The answers to those questions, I hoped, had the potential to improve the outcomes for people around me. This was thrilling. I wasn't hopping boxcars across the West, but in a way, I was becoming a version of the horse handicapper I'd fantasized about. Many of the statistical techniques my father had worked out for the fillies were similar in spirit to those I was learning in graduate school for the analysis of human data. His methods were inelegant, as had been my computer implementation of them. And our algorithm hadn't worked so well in California. But he had known to look for all the relevant variables in a system, as I was now doing.

All I needed to arrive at the answers I craved was a way to understand and identify the importance of each variable and the relationships between them. What I needed was the equivalent of my father's colored pens.

IT'S SAID THAT ALL OF SOCIOLOGY CAN BE BOILED DOWN TO EITHER OF two questions: How do groups hold together (or fall apart)? And, what makes groups unequal? My interests lay in the second category. The subfield called *social stratification* addresses the systematic differences in power, status, and economic resources that pervade every modern society. Some scholarship in this vein asks what makes some societies more or less unequal than others—that is, how far apart are the rungs on the economic ladder? I am not as interested in this aspect. Rather, I am fascinated by what is called *mobility research.* Namely, given the existence of an economic ladder, what determines who ends up on which rung—and, compared to one's parents, who climbs and who falls?

If you had asked me at the time, I would have said the answer had almost everything to do with nurture. My graduate dissertation, which was among the first scholarship focused on the racial wealth gap, used statistical models that, to use my father's lingo, exclusively factored in "training" and not "breeding"; in my research, I tried to determine how much parents' net worth explained who succeeded and who didn't and whether this accounted for Black-white gaps in education as well as social and economic outcomes. I got exciting results that seemed to show that parental wealth was, in fact, critical. The average Black family has only about a tenth of the net worth of the typical White family, and I found that this disparity in assets tracked closely with differences in children's outcomes such as how far they went in school, what kind of first job they landed, whether they were economically self-sufficient as adults, and how much wealth they, themselves, accumulated. That is, among Black kids and white kids who grew up in families with the same parental education and wealth levels, there was no difference in outcomes.

Until then, academics and policymakers had only taken income differences into consideration and had ignored the huge wealth gap when drawing conclusions about the difference in outcomes among racial groups. They simply hadn't known about the extent of the racial wealth gap nor about the power that parental wealth seemed to exert on offspring's life chances. My work was the first to show that the wealth gap was, in fact, hugely significant in perpetuating racial disparities, and I thought I had found one of the key linchpins of (dis)advantage across generations. In my thesis, I argued for a whole new kind of welfare policy, one that focused not just on *income* support but on building *assets* in poor communities.

But then I started to question my own claims. As I worked to turn that doctoral dissertation into a book, I read *What Money Can't Buy: Family Income and Children's Life Chances*, by Susan Mayer, a professor at the Harris School of Public Policy at the University of Chicago. Mayer argued that the way we had been assessing the effect of poverty was all wrong. Typically, researchers like me would line up survey respondents according to their race, educational level, age, and maybe a couple of other characteristics. Statistically factoring out the effects of all those variables, we could then see how people who were otherwise alike but had different income levels turned out. If, all else being equal, the poorer ones had worse health or children who floundered in school, we concluded that income differences caused those outcomes.

But Mayer suggested that this was myopic. She discovered that a dollar from a transfer payment (i.e., welfare or gifts or inheritances) had little to no effect on children's outcomes, whereas a dollar from earnings had a much bigger effect. If it was really money per se that helped children, why would it matter where the dollar came from? Maybe, she argued, there were "hidden factors" that had led parents to earn more in the first place, and those same hidden factors were actually responsible for improving their children's health, wellbeing, and educational performance. It wasn't the income itself that made a difference, in other words, but rather whatever factors had led to that income.

Moreover, when Mayer measured parental income in the years

after a child's cognitive ability was tested, when the money couldn't be influencing the test scores, it mattered almost as much as parents' income *before* the assessment did. Unless you know of some space-time wormhole I haven't heard of, a parent's future income cannot affect a child's current performance on a test. This evidence hinted at the possibility that it was not actual dollars that were driving kids' success or failure to thrive. Rather, there was something about parents who had higher incomes—some underlying attributes—that was advantageous. Parents who worked hard, who did well on tests, who were able to delay gratification, were conventionally attractive, or who had whatever else it took in our society to earn a lot of money also tended to be parents who passed on these adaptive traits to their offspring—culturally or even (gasp) biologically.

Mayer's book rocked my world. Her ideas upended the field of mobility research as far as I was concerned, lifting a veil on the problems inherent to the approach I had been using in my own work. Yes, she had been looking at income, and I was examining wealth. But the issues were similar.[2] Just as she had shown that the true, causal effect of income was probably overstated, the same could be true for parental net worth. Perhaps the parents' assets, per se, didn't drive a child's success either, but rather the underlying traits that had led the parents to accumulate the assets.

From good historical evidence I had reviewed, it was fairly clear that the effects of slavery, Jim Crow laws, redlining, segregation, and a host of other structural factors and policies were the primary drivers of racial wealth gaps. And, in turn, it was pretty obvious how wealth mattered, smoothing the way for offspring as they made their way into the world. Parental assets meant the difference between graduating college debt free or burdened by student loan payments (or not going to college at all). Family wealth meant being able to borrow the money for a down payment as a young adult buying one's first home. Or inheriting the family business. Still, like all good science should, Mayer challenged my assumptions and left me feeling unsettled, queasy even. There had to be a better way to assess the impact of social environment on children's outcomes—and I needed to find it. I wasn't the only one looking, as it turned out.

LITTLE DID I KNOW, BUT A WHOLE DOPPELGANGER FIELD WAS LURKING in the psychology building across campus: behavioral genetics (or BG as it is known). The goal of BG is to explain what shapes our behavioral, personality, and cognitive traits, and what the field has found is that people from the same family resemble each other not because of their shared experiences but rather their shared genes. Sociologists have assumed that when we see that high-income parents are more likely to beget rich children, that is due to social transmission within the family. BG shows that this is usually not the case: nature makes us the same as our sisters and brothers and parents; nurture makes us different from them.

Historically, BG has been able to point to how much genes and the environment each matter without being able to pinpoint specifically *which* genes or *what* aspects of the environment are key. In one type of behavioral genetics study, researchers will compare adopted children to both their biological and adoptive parents, respectively, to see, for various traits, which set of parents they most resemble. The use of adoptees to understand the relative impact of nature and nurture seems straightforward. To the extent adoptees resemble their biological parents, that must be genetics. And, assuming they are adopted at birth, to the degree they are like their adoptive parents, that must be the influence of the social environment. However, biological parents who give up their children for adoption at birth are not a random sample of the population. They are more likely to be younger, financially unstable, and so on. They are also more likely to be religious—hence, opting for adoption over abortion.

Meanwhile, the folks who adopt children are also not a random slice of the public. They come in two tranches. The first are relatives adopting children whose parents can't take care of them—aunts or uncles, grandparents, even older siblings. Such familial adoptions are necessarily excluded from the analysis, because the approach is predicated on the fact that the adopters and adoptees are genetically unrelated. But even among parents adopting unrelated babies, they are not random. First off, most adoptive parents are in the market for children, so to speak, because they have some medical issue that prevents them from having biological offspring. Moreover, they tend to be well-off, educated, and religious, on average. They are also people who tend to have a certain conception of family and love that is not based on blood ties. In short, they are not a representative slice of parents.

How all these issues affect estimates of genetic and environmental influence is not straightforward, however. Narrowing the range of environments to which adopted children are exposed (since adoptive parents do not represent the full extent of families) reduces the apparent influence of nurture. But the fact that the biological parents may be more alike genetically than random parents reduces the measured impact of nature. So perhaps it all evens out? Some studies managed to use adoption agencies that randomly assign kids to families. Others have compared siblings within families where one is a bio kid and the other is adopted. The results pretty much hold up.

A bigger, unresolvable issue with adoption studies is the fact that any resemblance to the birth parents may not just result from the genetic connection between bio parent and offspring but also could be due to prenatal environmental conditions—secondhand smoke, stress, medication, and so on. We know prenatal environment matters to outcomes, we just don't know how much, exactly, or what particular aspects matter the most, so I'm afraid adoptees will never be the perfect study design that behavioral geneticists want them to be.

In an alternative to adoption studies, behavioral geneticists turn to identical twins. When most people envision twin studies, they are imagining identical twins separated at birth and reared apart. Such cases are rare, but not unheard of. The movie *Three Identical Strangers* recounts the story of the Edward Galland, David Kellman, and Robert Shafran, identical triplets who had no clue of each other's existence until age nineteen, having been adopted out to three different families. Upon meeting, they discovered uncanny resemblances: their favorite brand of cigarettes, the way they sat, the fact that all of them wrestled in high school, just to name a few. I won't spoil the movie for you, but the logic of the film is the same as the logic of the Minnesota Study of Twins Reared Apart (MISTRA). If these triplets who didn't share the same environments displayed such eerie similarities, that must mean that genetics is a lot more important than most of us give it credit for.

But problems abound with these sorts of studies. Firstly, when the triplets find out that they like the same brand of cigarettes, that's amazing, but it's the same sort of amazing as when you are on a plane and end up chatting with the stranger in the seat next to you. After a few minutes of conversation, you might find out that you both love the Boston Red Sox even though neither of you hail from the Bay State. Or that you both love fly fishing or crochet or needlepoint. Finding similarities with other humans is a basic instinct in social animals such as ourselves. Dig deep enough and you might find commonalities with anyone. But the list of a dozen eerie similarities among the triplets is out of what denominator? In other words, whether it's your long-lost twin or the stranger on the plane, how many potential dimensions of similarity did you consciously or unconsciously scan through before landing on what clicks between you? The possibilities are literally endless.

So, to be certain that we are not being led by our own biases toward finding patterns of similarities, we need to predetermine what factors we are going to examine. This is called *preregistration of hypoth-*

eses in the world of psychological research. But even if we have a small and fixed number of preannounced dimensions on which we are going to examine twins reared apart—say extraversion, height, weight, and political ideology—there are still problems. For instance, in MIS-TRA, many of the twins weren't actually separated at birth. A third lived together for at least their entire first year of life. To the extent that early childhood environment matters the most (as many scholars think is the case), any similarities could be due to nurture as well as nature. (That's on top of the prenatal environmental issue raised earlier with respect to adoption studies.) Moreover, over three-quarters of the twins reared apart maintained contact with each other over the course of their lives. That's a leaky test tube they are being raised in, allowing for environmental cross-contamination. And more than half (56 percent) were raised by a close family member (so much for random environments), and almost a quarter ended up being raised together again at some point or living as next door neighbors.

As compelling as long-lost twin studies are, they are better for movie scripts than for science. In fact, most bread-and-butter twin studies from which we get estimates of nature and nurture are a lot less dramatic than reuniting long-lost twins. Instead, they focus on twin pairs who grew up together. How this sort of twin study works is fairly straightforward. Identical twins are, by definition, 100 percent genetically similar. Fraternal twins are, *on average*, 50 percent genetically identical, like any non-twin full siblings. (Studies restrict fraternal twins to those of the same sex to be comparable.) So if identical (monozygotic) twins resemble each other on, say, personality traits more than fraternal (dizygotic) twins do, we can chalk that difference up to the increased genetic likeness of the former.

By doubling the increased likeness of identical twins as compared to fraternal twins, we obtain the *heritability estimate*: the proportion of population variation in extraversion, height, education level, BMI, and so on that is due to genetic variation in the population. Meanwhile, 100 percent minus the heritability is the *environmental influence*. We can further break down this environmental influence into two components. The first is the influence of what behavioral geneticists called *shared* or *common* environmental influences. Having a shared or common environment means all aspects of the environment—air and water quality, socioeconomic status, neighborhood, parents, and so on—are experienced the same between siblings. And unique environment influences are those that are particular to each brother or sister: If one, by chance, had a great math teacher and the other didn't, for instance. If one was treated differently by the family than the other (as long as that differential treatment was not due to genetic differences). If one had a serious accident and missed a chunk of school in eighth grade. The key is that these environmental influences have to fall into the random events basket. In other words, they can't be environments that the siblings chose or evoked because of their genetic differences.

To get the estimate of unique environmental influences, one merely has to look at how similar identical twins are. Identical twins share their genes 100 percent as we already know, and, it is assumed, they also share their common environments 100 percent. So, any differences between them result from the unique events that happened to one and not the other.

This classic twin model is called the *ACE model. A* stands for additive heritability (that is, the heritability ignoring the fact that genes may have multiplicative effects on each other); *C* stands for common (i.e., shared) environment; and *E* stands for error—that is, unique environment. (I guess ACU didn't sound as cool as ACE.) Variations on this model try to estimate dominance effects (i.e., the multiplicative effects of one versus two copies of a given gene). And *extended twin models* incorporate other relatives like half-siblings, double-cousins (kids of a parent who is a twin), and so on.

The ACE model rests on two key assumptions, however. (Those

were the soft underbelly I would attack in trying to save the sociological, blank-slate way of looking at human differences.) The first
critical assumption is that fraternal twins are, in fact, 50 percent
genetically similar, on average, with respect to any of the relevant
genes for extraversion (or whatever the relevant phenotype is). This
will be true if parents mate at random with respect to those underlying genes. However, if parents with more extraverted genes tend to
reproduce with those who tend toward the introverted side, genetically speaking, then fraternal twins are going to be related at an average that is below 50 percent for the relevant genes. This would lead to
an overestimation of heritability.

As it turns out, whatever adages may say, it is more often the case
that like attracts like rather than a scenario of opposites attract, at least
at the genetic level of human mating. Tall people tend to marry tall
people. Smokers marry each other. And intellectually gifted men and
women tend to mate with each other. *Homophily* or *positive assortative
mating* (the two technical terms for this dynamic of like pairing with
like) results in the opposite dynamic whereby fraternal twins are more
genetically alike for the relevant markers than the models assume they
are. This, in turn, leads to an underestimation of heritability.

We can, in fact, know how alike parents are on a given trait. Now-
adays we can even measure their similarity on the genetics of those
traits thanks to reliable genetic information in some social scientific
surveys that include spouses. So, genetic assortative mating is something that can be tested for and taken into account. Indeed, my colleagues and I found that spouses tended to be correlated at plus 0.15
for education PGIs (a zero correlation would indicate random mating). For the genes related to height, the correlation was 0.31. Other
studies followed and found that the spousal correlation in education-
related genes in Norway, for example, was even higher: 0.37. These
results suggest that if anything, the goal posts of heritability were
closer than they should be, that heritability was underestimated.

But it was the other key assumption of twin models for which I was gunning, one called the *equal environments assumption* or *EEA* for short. Going back to the example of extraversion, we are assuming that monozygotic twins are more alike on extraversion *only* because they are genetically more similar and not because they tend to experience more similar environments by virtue of being members of an identical twin set. At first glance, it would seem a given that MZ twins experience more similar environments than do DZ twins.

But it's only certain types of environmental similarities that would mess with the equations. If identical twins experience more similar environments, such as acting classes or party-going or other group social activities, because they are more alike in their extraversion genetics, that's just an example of genetic effects working through the environment, as we will discuss in Chapter 6. But if they are treated more similarly solely by virtue of being identical twins, then this represents a violation of EEA.

The day before writing this, I was at the playground with my son. With me was one of the economists who is a leader in the sociogenomics (or, as he would call it, genoeconomics) field. We watched as two grandparents unloaded twin boys from a double stroller. We turned and glanced at each other knowingly because the toddlers were dressed identically. Same gray and red sneakers, matching gray and red-striped sweaters that seemed to go with the sneakers, and the same dark blue pants. *EEA violation!* we were both surely thinking. The three-year-old boys' genes had not led them to pick out the same clothes via genetically encoded taste. This was an environment being imposed on them by virtue of them being identical twins. The same is true when twins are confused for one another. Or when they share friends because people think of them as a package deal. Or if identical twins spend more time with each other and influence each other to a greater degree. All these causes of similarity have nothing to do with their genes but instead result from twins occupying a special role in our society.

Until the age of human molecular genetics, the equal environments assumption could not be tested. We could not know whether wearing the same kind of shoes or belonging to the same friend clique was more often due to twinness or to genetically informed taste. Moreover, we could not know if shoe-wearing or any other measure was an important aspect of the environment. We could measure how much time twins spend together, but again, we could not know how consequential that time spend is in terms of shaping how alike they become on a given outcome.

But with molecular genetic data, we now had a new possibility—a truly "natural" experiment. In the days before genotyping occurred, survey researchers had concocted a series of questions that they asked twins or their parents to ascertain their zygosity. They were along the lines of "Were you and your twin as alike as two peas in a pod?" "Were you and your twin mixed up as children?" and "By whom were you mixed up?" In the age of DNA testing, when we can verify true zygosity, we now know that a scale made up of responses to these three simple questions predicts actual zygosity 93 percent of the time. Meanwhile, just asking twins (or their parents) if they are identical or fraternal gets the right answer only 82 percent of the time.

It is within the 7 to 18 percent of twins who are wrong about their status that the natural experiment lies. Twins who have grown up under the wrong assumption—that they are fraternal when they are identical or vice versa—provide a perfect test of this hitherto untestable assumption of equal environments.[3] The more common error is for identical twins to think they are fraternal. I decided to use these twins—those identicals who have lived their whole lives thinking they are fraternal—to stand the equal environments assumption on its head. If I found that they were less similar on a range of outcomes than monozygotic twins who lived their whole lives as monozygotic twins, then it would raise the possibility that part of the reason that identical twins were more similar was the unique environmental quirks they share for being clones.[4] But finding that they were just as similar in GPA, BMI, ADHD, and so on would mean that whatever special nurture identical twins received that other siblings did not receive was inconsequential for heritability estimates. In other words, the EEA would be validated.

As it turned out, across three different datasets, the heritability estimates we obtained were the same, whether we used misclassified or correctly classified twins. In other words, these misclassified identical twins were as alike as correctly classified identical twins, suggesting that the regime of environmental similarity or dissimilarity they were living under didn't really matter. The EEA held. And I had to concede defeat in my last gasp as a pure social scientist doing "battle" against genetics.[5]

Despite the limitations inherent in adoption studies and notwithstanding the absurdity of twins-reared-apart surveys, the ACE model held up, and its heritability estimates were higher than I was comfortable with as a social scientist. There is no perfect equation in the human sciences, but the various approaches to estimating genetic effects tend to converge around the same robust answers. Namely, what psychologist Eric Turkheimer has called the first law of behavioral genetics appears to hold: all human behaviors are at least moderately heritable. Indeed, in 2015, Tinca Polderman and colleagues published a paper that looked across fifty years of twin studies. They examined 17,804 traits and found that across all traits that have been studied using this method, the average heritability—the impact of genetic variation on outcomes—was 49 percent.[6] Forty-nine percent! We are equal parts nature and nurture.

Nurture, in this case, includes both the environmental factors shared between siblings and the unique exposures that differ between them. Recurrent depressive episodes were 45 percent heritable (A), but only 3 percent of the variation was due to the common family environment (C). The quality of intimate relationships was 28 percent heritable and 6 percent attributable to C. Temperament and personality? Forty-four percent heritable and 13 percent common family environment. Memory (A = 46%; C = 6%); sleep disorders (51% vs. 12%); sensation of pain (40% vs. 3%); health behaviors (41% vs. 13%). The only outcome where common family environment mattered as much as or more than both genetics and unique environment was religious/spiritual beliefs (A=31% vs. C=35%). For all other measures, the common family environment was generally the least important component when compared to genetics and random environments.

I had overlooked the significant influence of pedigree in my work, just as the horse handicappers had done in my father's era. Twin research implied that genetics played a much bigger role in the lives of the children I studied than I had imagined. But even more staggering to me—a sociologist who cared about inequality—were the socioeconomic implications of these twin studies. When researchers bracket off the issue of racism by focusing on non-Hispanic whites, perhaps we find that white parents with higher incomes just passed more economically advantageous genes on to their kids than white lower income parents, and the money itself wasn't all that important to their kids' success? The evidence certainly allowed for that possibility. As far back as 1975, economists had been trying to estimate genetic influences on income using the same methods behavioral scientists had used for personality traits—by comparing identical and fraternal twins. Later studies using better samples concluded that variation in genes in the U.S. explained 58 percent of the variation in income in the male population and 46 percent in the female population. In other words, genetic differences explained about half of where we end up on the income ladder. As far downstream as labor market earnings may be from cells, proteins, and DNA, it seems that these genetic differences between people somehow translate into differences in economic success. That fact alone might explain why

Susan Mayer, the author of *What Money Can't Buy*, and others found such weak effects of actual income on children's outcomes.[7]

That is, genetic differences have long been known to have a large influence on variation in children's test scores. So, if the genetic influences for income and those for test scores overlap, which they likely do, then the fact that parents pass on their genes to their children must certainly inflate the apparent effect of income. Moderately high heritability estimates for income imply that we might have already had the answer to the question Susan Mayer raised: Could it be genetics, not culture, that was biasing the effect of family income on children's test scores?

(Of course, my own dissertation was about *wealth*, not income, which was not examined using twin studies until 2012, when it was found that wealth had a much lower heritability than income does—about 30 percent.[8] My own estimates using a different approach on new data confirmed this figure. Moreover, since twin studies are conducted within a given racial group, they imply nothing about race gaps, which are likely due to the legacy of slavery, Jim Crow, redlining, and a whole host of discriminatory practices that I documented in the book that emerged from my doctoral research. More recent scholarship also shows that wealth differences do not reflect genetic differences to the same extent that other measures of socioeconomic status do.[9] Though it may be counterintuitive, the very fact that much of wealth is literally passed down between generations makes it more socially [i.e., environmentally] determined as opposed to genetically influenced. The only part that we inherit biologically is the ability to make wealth. And, as we know, wealth involves a lot of luck, not just ability.)

Indeed, parental income probably had little impact on children's outcomes *at all*, since behavioral genetics had shown that the shared family environment—earnings, but also parental education, age, occupation, neighborhood—had a very limited impact within the normal range of variation in a given racial group.[10] The sociological paradigm seemed to be underinformed. Not only had sociologists just waved away the huge effects of genes, but we had also been studying the wrong part of the environment—family conditions—rather than influences that were *not* shared by family members, which mattered much more. We had simply focused on the wrong things. I was becoming a secret heretic within my own field.

The problem was that simply knowing that genes mattered a great deal for social and economic success didn't help me. So, genes and environment worked together. But how exactly? And to what degree? There was no way to measure the impact of genes directly: the twin studies merely told us how much genes and environment contributed to differences in the population; they didn't pinpoint the genes directly, and they didn't allow us to stage a head-to-head race between say, genes and parental income. Until we could measure the actual genes, we wouldn't be able to explicitly test their impact, or that of environmental factors, for that matter. Though I was buoyed by the low-heritability estimates for family wealth, until we measured the genes behind both wealth accumulation and children's educational outcomes, I wouldn't know how much of the parental wealth effect was real. I could see the mist from the white whale spouting over the horizon, but I couldn't yet get a bead on the creature itself.

BORGNY AND HARRY EGELAND MARRIED IN OSLO, NORWAY, IN 1923 AND had two children: Liv, a girl, born in 1927; and Dag, a boy, in 1930. By the time she reached school age, Liv was able to speak a few words, understood only by her parents, but she could not construct complete sentences. She was described as restless and fluttering, with a "canine" appetite. Her brother was even worse off. He couldn't sit upright without assistance. He could not speak, chew solid food, or even focus his eyes. The children gave off a strong odor that their father—who suffered from asthma—could barely tolerate. To try to help their children, the Egelands consulted not only various medical professionals but also herbalists and even a psychic.

Their desperation led them, in 1934, to consult a Norwegian biochemist, Asbjørn Følling, with whom Harry had taken a class as a dental student. Følling performed a routine physical examination of the children, which he deemed unremarkable. He then went to test their urine for ketones—a byproduct of the body burning fat rather than carbohydrates for energy—in order to check for metabolic or dietary problems. Normally, when ferric chloride is added to urine, it turns red-brown if there are no ketones and purple if there are. In the case of the Egeland children, it turned dark green for a few minutes before fading. Følling knew that aspirin and other medicines could mess with the results, so he asked their mother to refrain from giving the children any non-food substances and to return a week later. Again, the samples turned green. He continued to analyze urine specimens from the children collected every two days by their poor mother (Dag was not toilet trained)—a total of forty liters of pee.

After weeks of testing, Følling discovered the compound phenylpyruvic acid in their urine, which accounted for it turning green. Følling suspected that the phenylpyruvic acid was related to their condition, so he went on to test the urine of four hundred other intellectually disabled individuals in and around Oslo. Eight of those subjects turned out to have phenylpyruvic acid in their samples— two sibling pairs were among the eight. Besides the acid in their urine, Følling observed that these individuals shared other characteristics: "fair complexions, eczema, broad shoulders, stooping figure, and spastic gait. All suffered from severe intellectual impairment as well."[11]

In a 1934 paper he published with these results, Følling hypothesized (correctly, as it turned out) that the phenylpyruvic acid resulted from an inability to metabolize the amino acid phenylalanine. To test this theory, he needed to see if there was excess phenylalanine in the patients' blood. Using the bacterium *Proteus vulgaris*—which converted the amino acid to phenylpyruvic acid in vitro—Følling found that there was indeed too much phenylalanine in the blood samples.[12]

Though it wasn't until decades later that the specific gene for the hereditary disease—by then called phenylketonuria (or PKU)—was discovered, in 1938, Følling correctly deduced that it was recessive (only expressed when a person has two copies) and that even carriers (those with one copy of an ineffective gene) had slight amounts of phenylpyruvic acid in their urine. He had obtained a sample of phenylalanine—which was, at the time, very expensive and difficult to come by. Noting that he, too, was fair of complexion and suffered from eczema, he ingested a large dose himself and discovered phenylpyruvic acid in his urine. He later systematically tested carriers (parents of afflicted children) and found that they were typically much less effective at metabolizing phenylalanine as evidenced by the acid in their urine.

The 1986 discovery of the phenylalanine hydroxylase (PAH) gene on chromosome 12 represented—along with the pinpointing of the sickle cell mutation in 1959 and the gene for Huntington's disease in 1983—one of the most canonical examples of how hereditary disease could be located in specific genes. And this discovery helped launch a bigger quest: to pinpoint the genes for *all* things. In 1990, the Human Genome Project, the largest scientific collaboration the world had ever seen, was launched to map the entirety of the human genome. It was only a matter of time, scientists believed, before the genes for heart disease, schizophrenia, and even IQ would be discovered—opening the way to life-enhancing interventions. President Bill Clinton captured this hope in June 2000: "Today we are learning the language in which God created life," he declared, announcing the imminent unveiling of the first draft of the DNA sequence. "With this profound new knowledge, humankind is on the verge of gaining immense, new power to heal."[13]

When the final draft sequence was unveiled to the public in 2003, the first big surprise was how many—or rather how few—genes humans had. Many scientists involved in the Human Genome Project had been part of a betting pool, and nobody even came close to the right answer: twenty thousand. Even a lowly corn plant has one hundred thousand genes. (It turns out that there is no clear relationship between gene number and organismal complexity.) But the hope of finding the answers to human flourishing in our DNA code appeared well-founded. Scientists located important genes—such as the ones that contribute to Alzheimer's disease (APOE4) and breast cancer (BRCA1 and BRCA2)—and pinned down the mutations responsible for major diseases such as Tay Sachs, cystic fibrosis, and sickle cell anemia. I was enthralled by these revelations—so much so that in 2007, I decided to pursue graduate studies in biology so that I too could learn the language of DNA, but in my case, as it related to human behavior. The great white whale seemed to be almost in reach—with one more big gust of wind directed at my intellectual sails, it might come into range.

AS I WAS PURSUING MY BIOLOGY PHD AND RAISING TWO CHILDREN, I learned that my first wife was leaving me. She had always been an ambitious artist, but as her international renown grew, she got more and more commissions that took her to farther and farther corners of the globe. She often invited us to accompany her, but the kids had school, and I had work, and the idea of toting two kids on red-eye flights and watching them in a foreign city by myself while she was working held little appeal. So, mostly, I held down the fort while she left to give talks and install shows. She began spending more and more time in London, and decided, ultimately, to live bicontinentally.

Elaborate rites and rituals are associated with the start of a marriage, but few mark the end of one. I wanted to mark this ending, I decided, and also to mark the beginning of my *own* new trajectory into the study of genetics. I would get a tattoo. But what would be the subject?

Since before the Human Genome Project, scientists had used experiments on mice, analysis of family pedigrees, and other techniques to zero in on genes of interest—kind of like persons of interest in a criminal investigation. They called these *candidate genes*. Many were for diseases, but some were for behavior. Around the time my marriage was unraveling, I was studying a gene that was reputed to be involved in impulse control—the dopamine receptor 2, or DRD2. (We have multiple dopamine receptors.) In mouse experiments, this gene had been shown to influence novelty seeking and risk-taking behavior. When compared to another version, one version of this gene, located in chromosome 11, seemed to result in a higher risk of alcohol dependency, binge eating, ADHD, antisocial behavior disorders, and other psychiatric problems.[14]

Curious, I decided to check my family's files. When I started graduate school and became fascinated with genes, I'd had the DNA analyzed for everyone in my circle—wife, kids, parents, sister, brother-in-law, nephews, even my kids' babysitter—with their permission, of course. Now, looking through the charts, I found that I had two copies of the safe version of the gene, as did both my kids. I had found the subject of my tattoo. I decided I would etch into my skin the DNA sequence of the "cautious" allele of DRD2, the version my children and I shared. What better way to mark the end of my marriage, while also exhorting the future me to behave in accordance with my own genotype? After all, despite having the safer gene, I had jumped into marriage and fatherhood too quickly.

A couple of hours of painful needling later, eighteen letters for the top strand—CTGGTCTAAGTCCATGAT—were emblazoned across the inside of my forearm, along with the complementary strand of nucleotides (the other stairwell on the double helix), upside-down and right to left: ATCATGGACTTAGACCAG. The entire gene consisted of thousands of bases; I didn't want my whole body covered, so I stuck with the region right around the mutation in question. The tattoo artist kept saying how cool my design was. He had done a few double helices in his career but never an actual DNA sequence. My pride in the thirty-six letters ballooned when my tatted-up brother-in-law oohed and aahed over the design and when a graphic artist friend asked if she could photograph my arm for the next edition of her book, *Body Type*. I was marking the end of one chapter, I thought, but also the beginning of another, an exciting intellectual period in which I was getting closer to the answers—now that we could measure genetic inputs—that I had been seeking for a decade. The tattoo would serve as a permanent reminder of this pivot point, a coded instruction to move forward and not regress in my life.

Soon, however, I was plagued with doubts. The tattoo was low on my arm, and I felt uncomfortable in formal work settings with it visible. Much more disappointing were some revelations about the gene itself. New studies with better samples and improved statistical designs could not replicate the initial findings about the gene's effects on impulse control or risky behavior. What had symbolized a death and a rebirth, an end and a beginning—a certain *logic* for my life—was maybe nothing more than a random bunch of letters of no consequence. The tattoo became a reminder, instead, that all (scientific) knowledge is tentative and subject to revision.

The failure to replicate the DRD2 results was part of a huge shift in human biology: from the study of candidate genes to the study of the collective effects of many genes, an approach known as *genomics*. Instead of the effects of a given gene, genomics describes the *genetic architecture* of traits. Think of our DNA as 3.1 billion beads along a string; each bead can be one of four colors. Only about 0.1 percent, or one in a thousand beads on that string that makes up our genome, commonly differs across people.[15] It took a painful decade of research to learn that most traits and diseases were actually influenced not by a few key genes but by *thousands* of tiny genetic differences—effects much too small to be detected in the puny samples on which we were running candidate gene studies. The OGOD paradigm of human genetic effects—one gene, one disease—bit the dust for many outcomes we care about.

In those early days, when we first realized how polygenic most traits or diseases are, this came as an incredible disappointment. At first, geneticists despaired: How could they ever understand the fundamental biology of, say, high blood pressure when it involved hundreds of genes? What hope was there of devising a pill to combat a devastating, heritable condition like schizophrenia if that medication needed to mimic or block a thousand genes to do so—and those thousand genes were implicated in many other biological processes in the body? Moreover, genotyping or genome sequencing was so expensive that there was little hope of gathering enough samples to detect the subtle effects spread across all the chromosomes. This kind of analysis required genomic data from hundreds, thousands, or even millions of people to detect a potential effect for each of the beads on the string of life. It was hard enough to sequence a single genome—the Human Genome Project alone had required thirteen years and 2.7 billion dollars.

Luckily, however, the cost of genotyping was falling—and fast. Thanks to next-generation sequencing methods that chopped DNA into smaller pieces and lined them up against a reference genome, the cost of sequencing a genome began dropping more quickly than even the cost of microcomputing—which was by about half every eighteen months. It was suddenly cheaper to analyze someone's DNA than it was to hire someone to ask them a bunch of survey questions. Today, a whole genome can be read—all 3.1 billion base pairs—for less than a hundred dollars. *Genotyping*—a shortcut that gives us information on only the three million sites we know commonly show variation between people—is about twenty-five dollars a sample.

For researchers, this affordability meant that we could suddenly abandon focus on a handful of curated genes and throw the genetic spaghetti against the wall, so to speak, and see what stuck. This hypothesis-free, comprehensive approach was transformative. Instead of geneticists being like the proverbial drunkard at night—who, when asked why he is only looking under the lamppost for his lost keys, replies, "Because that's where the light is"—we could finally find the keys we'd been searching for because we could look in areas that had once been scientific darkness. The advent of cheap genotyping was like moving from a little sauce pan to a giant pasta pot; it could handle *all* of a human's DNA.

The spaghetti tossing worked like this: There are four colors for beads (nucleotides). We can "score" the variations. In what is called a *genome-wide association study*, or *GWAS*, we can now track these variations among all these beads for as many people's DNA as we have in our sample. Let's say, for simplicity, that we find that the people with one variation are taller by 0.1 inches, on average, than the people with the other variant at a particular location on chromosome 1. We note that down. We then go on to the next bead. This time, we find the scored allele to be associated with lower height, by 0.04 inches. After we collect these results at each of the three million or so locations we've measured in each person's DNA, we can then take these effects and combine them. Some researchers only sum the biggest effects. Others add together the results for almost each of the three million beads. In general, the more beads you add together, the better the resulting index predicts the trait.

Once we have this index—that is, the polygenic index (PGI)—for each person in our sample, we can then see how well it predicts a person's height. The index might explain, for instance, 50 percent of the variation in height. This figure falls well short of what we know to be the full genetic influence on variation in height, which is thought to be around 80 to 90 percent in modern societies. But the fact that we can even get that close in prediction is astounding. Genome-wide scores have led to a much-higher caliber of research that consistently replicates across samples, labs, and publications. If my father had the same level of predictive accuracy for horses as PGIs offer for humans, he might have died a rich man. (Height is the best predicted outcome as of present; for others we can predict as little as 1 percent of the variation. Most traits are somewhere in between.)

Today there are PGIs for height, diabetes, BMI, schizophrenia, and even for cognitive ability and educational attainment. Take education: someone with an education PGI in the bottom fifth has a 12 percent chance of completing a four-year college degree, while someone in the top fifth enjoys a five-times-greater likelihood of getting a bachelor degree. Once we have big enough samples to run good GWASs, the only limitation is the degree to which outcomes depend on genes. In short, if you can measure it, you can calculate a PGI for it. The sequencing of the human genome has not yet led to customized pharmaceuticals to make us taller, leaner, smarter, and healthier. But it has led to a new science of prediction. Today, we can generate a (noisy) prediction of a U.S. child's adult height, how far the child will go in school, and whether that child will be overweight as an adult— all from a cheek swab, finger prick, or vial of saliva.

As with any new technology, there are some big caveats to this dawning age of genetic prediction. First, while the scores predict averages extremely well, they are far from deterministic. While 12 percent of people from the bottom quintile of the education PGI graduate college and 60 percent of those in the top quintile do, that still means that those 12 percent get more education than 40 percent of the top PGI-scoring kids—and that's comparing the extremes of the distribution, never mind the middle. The point is that while it is a breakthrough to predict at all from DNA, in terms of telling us how much school a particular individual will complete, the PGI is super noisy. Even when the indices improve and they capture the full heritability of a trait like education, plenty of low-PGI kids will complete doctorates, and plenty with ninety-ninth percentile scorers will drop out of school. But then, who would want to live in a dystopia where all of our futures could be assayed at birth (or before)? This limit to prediction doesn't mean that PGIs will be of no use. After all, we can't predict the weather perfectly, but we still rely on meteorologists.

A second caveat worth noting is that PGIs have been developed and work within only a given ancestral population. Scores that are trained on samples of people of exclusively European descent predict best among those same people. Their accuracy declines when the PGIs are aimed at Asians, Africans, or those with indigenous American ancestry. There is tons of research going on right now to try to determine why this is the case—whether this "portability problem" is due to genetic factors, such as different distributions of the relevant alleles in each group, or whether it is due to social factors, like the different environmental landscapes that ancestral groups face. The bottom line is that comparing PGIs across groups tells us nothing about genetic differences between those groups—despite the efforts of ideologues like Charles Murray, in his book *Human Diversity*, to assert that they do.

The kicker is that since rich countries tend to have the money to fund genetic research, most of the data available are for people of European descent. So, we have the best-functioning scores for Europeans (and people of exclusively European descent living in other parts of the world like white Australians, Canadians, New Zealanders, and Americans). That means to the extent that PGIs are useful, they are useful for people of European descent, thereby creating new technological disparities. (I will address all these issues in Chapter 8, when I turn to the clinical and policy implications of sociogenomics.)

Just as the rich got VCRs, mobile phones, and just about any other technology first before they trickled down to the rest of the population, so it is with genomic science.

IN 2017, SOCIAL SCIENTISTS WERE STILL SEARCHING FOR WHAT SOCIAL factors made rich and poor kids turn out differently when a Princeton professor named Matthew Salganik announced a competition he called the Fragile Families Challenge. Salganik is a computational social scientist—that is, someone who addresses sociological questions with cutting-edge machine learning approaches. He had been reading about the Netflix Prize, a competition sponsored by the company that invited teams to play with some of its data to develop recommendation models. The winning team was awarded one million dollars. A light bulb went off in his head: If mass scientific competitions could help a private company recommend the perfect indie comedy, why not use the same technology to improve children's lives?

He went upstairs to knock on our colleague Sara McLanahan's door. McLanahan was the principal investigator of the Fragile Families (FF) dataset, the fruits of a survey that has followed ten thousand families for almost three decades, collecting thousands of variables over time—from a child's birth weight and housing conditions to parental employment and the quality of the child's schools. Salganik proposed his idea; she took him up on it, and the Fragile Families Challenge was born. The goal of the competition was to have teams from think tanks and universities apply machine learning and other novel computational models to see who could best predict from the variables captured from birth to age nine how kids' lives had turned out at age fifteen—how well they did in school, whether they experienced poverty and/or eviction, and so on.

In response to the call, 160 teams signed up for the challenge.

Participants hailed from institutions far and wide, from Northeastern University in Boston to the University of Lincoln in the UK. They came from across the social sciences—sociology, economics, and political science—but also from computer science and other technical fields. Even a plasma physicist entered. There's generally not a lot of real-time excitement in social science, so the idea of a real academic horse race was appealing, even if the grand prize was just an all-expense-paid trip to Princeton, New Jersey, to present the winning formula. It was also exciting from a public policy perspective. Machine learning was revolutionizing everything from internet searches to voice and image recognition. Why not unleash its immense power on some of the best data collected to identify which children were beating the odds and to better understand why?

The goal was to predict six outcomes among the FF children at age fifteen based on data from ages zero to nine: GPA, child grit (indomitable spirit; pluck), household eviction, material hardship, caregiver unemployment, and caregiver participation in job training. The FF Challenge wasn't like a chess tournament, the Indy 500, or the Kentucky Derby, where teams gather in one place for a few hours. The competition took place over weeks, and the teams merely uploaded their predictive models at two stages—an initial "lap" on May 10, and their final, tweaked approaches on August 1. As in the Indy 500, a leaderboard added to the excitement.

The organizers let the teams practice on about half of the data to develop their statistical approaches. When they submitted these, they were tested on leaderboard data that gave feedback to them about how well they had predicted the six outcomes. At the end of the competition, the final algorithms were tested on the other half of the data. In other words, the teams had access to information on some of the families to work out their best predicting algorithm, but that algorithm then had to predict children's outcomes in family datasets they had not had access to. The winners would be the teams who could predict the six outcomes at age fifteen the most accurately in the data they had not seen.

The results were ... utterly disappointing. Despite the state-of-the-art technology, the algorithms were lousy at predicting life outcomes for adolescents. Even the team that won could only explain 5 percent of the variation in how kids turned out. Salganik considered the hackathon an abject failure. In postmortem analysis papers, researchers speculated on what they might have missed. Was it the sample size? Datasets to train Google to recognize a cat have millions of entries; Fragile Families had only a few thousand. Or maybe the Fragile Families survey measured the wrong factors in children's lives?

This latter concern led to a whole new follow-up study of the Fragile Families subjects. Princeton researchers zeroed in on pairs of kids who were predicted by the best models to have the same outcome at age fifteen—say, an equal GPA—but who experienced different results (one had a C average and another an A average). They sent research assistants to those families to interview the members in open-ended fashion to see what—with no preconceptions—might explain those differences. They concluded that, most likely, the factors that had altered children's trajectories were random events like accidents or happenstance opportunities. Even though FF had collected fifteen thousand or so bits of data on each kid's experiences—the typical aspects of family, neighborhoods, and schools—these evidently didn't capture the environmental forces that ultimately mattered. Moreover, those forces were so idiosyncratic that there was perhaps little hope of ever systematizing them into a reproducible scientific endeavor.

Had we sociologists paid more attention to the disciplines around us, we might not have been so shocked by the results of the Fragile Families Challenge. Studies of identical and fraternal twins had shown for fifty years that the family environment that parents create for their children contributes very little to how those children turn out on a range of outcomes. That's not to claim that race or gender don't have huge environmental impacts on how we turn out. (Twin studies are almost invariably conducted within racial groups, so they say nothing about cross-race differences in environments; gender differences are not addressable by the methods either.) That's also not to say that food, clothing, shelter, and a loving home don't matter. There is no doubt that extreme deprivation, and its counterpart, enormous wealth, have environmental effects on how people turn out; most people at those extremes don't end up participating in these studies. More importantly, most of us are not at those extremes; we are somewhere in the middle of the distribution of environments, and there, the differences that parents tend to obsess over—this school district or that one—matter little. For instance, the shared household and neighborhood environments of siblings explains only 10 percent of our cognitive ability and basically zero percent of our eventual income. When it comes to most outcomes, twin research revealed that what matters most is not the environmental influences shared between brothers and sisters but rather what is *unique*, or unshared, within the family—such as events outside the household or differential treatment by parents. To treat the family as a one-size model was wrong. A family was not a printing press but a complex network.

For almost a century, we social scientists have been studying what's shared within families to explain how people turn out, but the environmental factors that explain differences among people within given racial groups in a particular society, it seems, are ones we have not been measuring. And since we have been studying the wrong things, we literally have no idea what those factors may be. Here's a back-of-the-envelope list of possibilities:

- Random accidents or infections
- In utero conditions (proxied by birthweight)
- Which teachers a child gets in school

- Who the child sits next to in class
- Birth order and gender combination
- The timing of the business cycle at graduation
- The timing of parental immigration
- The timing of parental death or divorce
- Whether a person was born just before or just after lead was removed from gasoline, iodine was added to salt, or fluoride was added to water
- Whether or not they were on the noisy side of the school
- Whether or not they were drafted
- Whether or not their military unit was deployed to combat
- Whether their grandparents were alive for a lot of their childhood
- How big the birth cohort was the year they were born

This is just my list; it's no better than anyone else's since, like the FF interviewers, I am starting from scratch—just brainstorming. Thanks to behavioral genetics, we know that the heavy emphasis of sociologists on what's shared about the environment among kids is very incomplete. Of course, besides studying non-shared environments, we should also study genes. As a matter of fact, thanks to the foresight of Sara McLanahan and her collaborator Daniel Notterman, the Fragile Families dataset (now renamed the Future of Families and Child Wellbeing Study) collected DNA samples from all the kids and their mothers. Sara McLanahan passed away in 2022, but I am in the process of trying to convince Salganik and the new director of the study, Kathy Edin, to run the FF Challenge V2.0: letting the teams have a crack at both the genetic and the social data, together. So far, I have not prevailed.

AFTER FØLLING CORRECTLY DEDUCED THAT PKU WAS A RECESSIVE hereditary disease, most medical professionals gave up on the hope for a cure. Informed by the binary thinking at the time, the widespread assumption was that any genetic disease is untreatable through environmental modifications. This is still true for some conditions, like Huntington's disease. But in the 1950s, British chemist Louis Woolf challenged this conventional wisdom, coming up with the radical—if logical—idea of treating PKU patients with a restricted diet that tried to minimize the amount of phenylalanine consumed.[16] Since PKU was a disease that meant the patient could not metabolize phenylalanine properly, removing it from their diets made sense.

Despite skepticism from his supervisors and colleagues, Woolf prescribed his dietary treatment for a seventeen-month-old PKU patient named Sheila under the care of Horst Bickel in Birmingham, UK. Within a few months of being treated with this restricted diet, Sheila learned to crawl, stand, and clamber up onto chairs. Her eczema disappeared, her hair darkened, and that smell that Harry Egeland couldn't stand in his own children had dissipated. When phenylalanine was added back into her diet—an experiment that would be considered unethical today—Sheila regressed.

Formal cognitive testing confirmed the miraculous results. Changing the environment had completely altered the downstream effect of the mutation, rendering it relatively harmless. In 1955, Wolf published the results of his experiments. By 1959, the first official PKU early detection and treatment program was in operation, in Cardiff, Wales. By the 1960s, screening programs were spreading across the world. In 1965, thirty-two U.S. states had instituted them; by the mid-1970s, newborn screening was routine across the industrialized world and in many poor countries as well. Today, newborns are screened for the disease in all fifty states, and you can read a warning on diet soda can labels and other food packaging—"Contains phenylalanine"—so those with PKU can avoid it.[17]

The idea that you could, in fact, alter the outcome of a hereditary disease through an environmental intervention was an early hint of an essential idea—that nature and nurture influence each other and, more specifically, that the effects of genes can depend strongly on the environment. It had seemed clear that this disease was entirely determined by genetics, as Følling had deduced. When it came to PKU, the nature-versus-nurture debate had seemed entirely settled in 1934. But somehow, twenty years later, Woolf had found a way to treat the disease *without changing the genes*. Suddenly, what we thought we knew about the categories of nature and nurture was not so clear.

The good news about this interplay between genes and environment is that diseases like PKU are no longer seen in such a fatalistic way. But it also means that this complex interplay makes it really hard to predict the courses of our lives. Let's return to the case of my childhood friends Leila and Crystal, by way of an illustrative, if speculative, example: Leila's fear of going outside is likely a direct result of both the neighborhood conditions (nurture) and the fact that she was already genetically primed to be anxious (nature). But even in Crystal's case, we may see the influence of both genes and the environment: perhaps a fascination with animal science on Crystal's part was also due to a genetic predisposition that transformed her desire to escape the neighborhood into a career that would take her to more pastoral environs, rather than cause her to retreat into her bedroom like her sister did. In fact, Crystal was the first in her family to move out of the neighborhood—to college, but then to Texas to work for the USDA, sparking her mother and sister to follow.

Sociology, and social science more generally, had been focusing on nurture, but now the field has been forced to concede that nature plays a major role in affecting the outcomes we typically study. At first, it seemed that all we had to do was try to factor our nature with our PGIs in order to get a truer sense of the impact of nurture. But in the decades since the PKU example, the field is now slowly coming to terms with that fact that including genes in our studies of the social world is not just about parsing out nature, it's critical for studying *how nature and nurture interplay.* In the case of Leila and Crystal, we would be blindfolded if we didn't consider both their genes and the environmental impact of our neighborhood.

The good news is that the PGI is not only useful for providing a measure of nature that was hitherto obscured; it also serves as an essential tool for understanding how genes and the environment intertwine to affect human outcomes. In this manner, the advent of polygenic prediction for social and behavioral traits represents nothing less than a revolution in the human sciences.

For the longest time, scientists studying human health and behavior had tools to measure the environment—lead levels in water, particulate matter in the air, the number of words a child was exposed to in early childhood, the level of bias in one's community, the quality of schools, and so on. But there were no equivalent metrics to assess the contribution of nature. Sure, we could study twins and adoptees to see how similar or divergent their lives turned out, but that didn't give us a measure of DNA that we could use in any meaningful way. Twin studies called to mind Plato's allegory of the cave, where we researchers could see the shadows cast by genes and the environment on the walls of the grotto but could never measure the things themselves. Wherein twin or adoption studies could merely infer the impact of genetics and the environment based on the metaphorical shadows cast by the similarity and differences among sets of relatives, PGIs provide us with a genetic X-ray of people, telling researchers about their DNA in useful and predictive ways. For the first time in history, scholars could truly integrate the biological and human social sciences. The great white whale was in sight.

The PGI has quickly become a critical tool for economists, demographers, and sociologists to answer previously vexing questions about, ironically, the social environment and how genes and the environment are mutually dependent on each other for their effects. Specifically, three aspects of genetic data distinguish them from other "big data," making them a unique implement for answering important questions in the human sciences.

First, it turns out that almost every trait that has been studied is at least partially influenced by that 0.1 percent of DNA that varies across people. Dimensions of personality such as openness or conscientiousness are only mildly influenced by genetic differences, but something like 90 percent of differences in the efficacy of working memory is thought to be explained by variation in that 0.1 percent of our genome. The genetic contribution to variation in how far someone goes in school is about 40 percent. For income, it's 70 percent. For cognitive ability, 75 percent. And for BMI, it's about 50 percent.[18]

Second, one's DNA blueprint is fixed at conception. Since DNA remains constant, it can help sort out causal directionality—namely, while my genes can affect your behavior (through my behavior), the reverse is not possible: your behavior can't change my genes. This means that we can use my DNA as a starting point to follow the arrow of how I may influence you and know that we are not suffering from some form of reverse causation (wrongly inferring that education causes good health when it's good health that causes education, for instance)—a problem that is rife in the social and behavioral sciences.

This immutable nature is what makes DNA distinct from other forms of big data that are washing over the social sciences like a tsunami. X feeds can be studied to better understand how polarized we are becoming. Facebook likes can tell us about contagion through a social network. And what people search for on Google when they think they are basking in the private glow of their laptops in their bedroom late at night informs scientists about people's true preferences on controversial issues like sex and race relations. But all these exciting new forms of data help scientists better capture *effects*. If we want to know about sexual orientation, we will do a lot better measuring searches for gay porn versus straight porn than asking people.

Genetic data are different. They help us better identify causes rather than improve our measurement of effects (i.e., outcomes). In this way, they are distinct not only from social media data but even from other biomarker data, which also typically capture outcomes rather than causes. A *biomarker* is any measure that attempts to capture biological processes or outcomes. Blood hormone levels could be a biomarker; the lipid panel your primary care physician runs for you at your annual physical constitutes a suite of biomarkers; even your weight-for-height and resting pulse are biomarkers. All of these measure outcomes, even if some indicate underlying risk factors.[19]

Finally, thanks to the magic of sexual reproduction, our DNA is the result of a random shuffle of our parents' chromosomes, making it the ultimate natural experiment. If it sounds like a dealer in Las Vegas is determining much of who we are and who we become, well, that's because our genetic makeup is a lot like a poker dealer mixing up forty-four decks of cards (one for each parental chromosome, excluding the sex chromosomes) for each parent.

By way of example, after getting my entire extended family to spit into vials, a genetic testing service showed that my daughter shares 27 percent of her DNA with my late father and only 23 percent with her grandmother, while my elder son is genetically more like his grandmother with exactly the flipped percentages. Because of these differences (and those on their mother's side), they are related less-than-average for siblings (44 percent identical rather than 50 percent). (That might explain why they seem so different.) Such variation from 50 percent is possible because I carry one copy of chromosomes from their grandmother and one from their grandfather. What each kid gets from me is a unique, random mosaic of those two grandparental chromosomes as they mix and meld into a single copy of each of the twenty-two autosomal chromosomes I pass on. So, if we dig down and examine my children on a chromosome-by-chromosome basis, we might find that on chromosome 1, my daughter is more like my mother but on chromosomes 2 and 3 she is more like my father, and so on. The sex chromosomes are different; I pass on the same Y from my father to both my sons. And to my daughter, I pass on an X from my mother. In this way, our genome is not like a single experiment with one blue pill and one red one. It's like thousands of randomized controlled trials (RCTs) bundled into a single person.

When deployed with PGIs, these three unique qualities of DNA allow us to study not just nature with more precision, but nurture, too, and most importantly, how the two are mutually dependent. That is, PGIs will not only allow us to better predict outcomes based on genetics, but they will also clarify how much different aspects of the environment matter. We have a new approach where we can factor out the genetics and see what "real" influence environments have. For instance, we can ask how much the effect of, say, parents' income on kids' educational progress changes when we factor out the parents' and the child's PGI for income and education, respectively. We can ask whether people at higher propensity for addiction to nicotine react differently to cigarette taxes than those with a low genetic propensity, even if they are in the same environment. Likewise, knowing how much our diet affects our cholesterol, net of genes, should inform medical practice—like how quickly doctors prescribe statins in lieu of lifestyle modifications. In other words, as in the case of PKU, measuring the genetics can guide us to environmental solutions.

But those applications are just the beginning. Deploying PGIs allows social scientists to study a range of topics that were hitherto unreachable—like the particle accelerator allowed us to examine quantum dynamics inside an atom—once we recognize that different PGIs not only react to environments in unique ways, but that, as in the case of Leila and Crystal, the PGI of one person can serve as an essential aspect of the environment of another. Namely, Crystal's genotypes—for education, for ambition, and so on—indirectly affected the environments of her mother and sister as well by causing them to move with her, blurring the notion that genes and environments are distinct causes on either side of some great ledger.

Do parents raise their children differently based on the innate characteristics of each offspring—that is, play favorites? Do the genes of a child provoke a certain environmental response from their parents—for example, does a pro-ADHD genotype in a child evince marital conflict among that child's parents? Driven by their genes, do peers influence each other's behavior to become more alike, or do birds-of-a-(genetic)-feather merely flock together? Is an elite college education a necessary environment for the translation of innate talent into economic success, or are schools merely good at picking genetic "winners" who would have done fine without four years of dorm food? Are bald men discriminated against by employers based on their unlucky genotypes? These are just some of the questions that have bedeviled the behavioral scientists lacking the proper tool to study them. It is as if it is the year 1600, and Zacharias Janssen has just invented the microscope.[20]

Today, I am a self-proclaimed biosociologist who seeks to integrate genetics and the social world. If, upon coming to the scene of a car accident, the sociologist notes that traffic fatalities have actually been declining, the biosociologist notes this and asks whether the declining pool of people who crash their cars is a random subsection of the population or whether we can predict who is likely to wrap their vehicle around a tree based on their genetic profile for alcohol consumption, risk-aversion, education, and so on. The biosociologist may even go on to ask how the genes of others may have played a role, perhaps those of other drivers on the road, or maybe those of someone in the passenger seat who was urging the driver to go faster. Finally, this new type of scientist also asks how policies around road safety evince different responses among people with different dispositions and may cause inequalities in risk to emerge.

In short, as a biosociologist, I measure people's genes as well as their environments and see how genes sort us into environments, how the impact of environment depends on our genes, and even how the genes of those around us matter for who we become. In essence, I am trying to gain a complete picture of how individual choices, random events, race (as opposed to ancestry), class, family, and neighborhood all work together with our biology to explain who gets ahead and who falls behind. This is known as sociogenomics, a field that I helped to invent, in much the same way I made up my own college major—this time with a dash of sociology, a sprinkle of econometrics, and a heavy dose of genetics.

The debate between nature and nurture, it turns out, is hardly as interesting as studying how nature and nurture *work together.* While I was once obsessed with factoring out genes to get a purer assessment of the environment, the rise of PGIs has completely reoriented my conception of who we are and how we thrive. It was hardly just the dry air in California that affected the horses at Golden Gate Fields. And it was hardly one gene variation that led to the end of my first marriage. The question of who succeeds socioeconomically is much more difficult to answer than I could have ever anticipated. But it's also much more profound, reaching as it does now into not just the social circumstances of a given time and place but into the very DNA that makes us who we are.

4

The PGI X-Ray Machine

In 1948, scientists began one of the nation's most successful long-term health studies.[1] At the time, heart disease caused roughly half of all U.S. deaths, and yet very little was known about its causes or potential treatments.[2] Back then, the study design was radical: rather than focusing on a select sample of people suffering from heart disease and comparing them with a control group of people without known heart disease, they picked a small town (Framingham, Massachusetts, because of its proximity to Harvard Medical School) and started gathering all sorts of information on as many of the adult residents as they could through physical exams, interviews, and specimen collection.[3] Moreover, the study would not just follow individuals; it would track entire families over the years. The Framingham Heart Study was the first long-term health study of its kind.[4]

The initial payoff to the Framingham Heart Study (FHS) came nine years later, when its first paper was published. It showed that high blood pressure was associated with an almost four times greater risk for coronary heart disease.[5] Another study soon followed showing high blood pressure to be a major risk factor for ischemic stroke as well.[6] Later studies uncovered other risk factors such as circulating cholesterol levels. Medicine in the twentieth century had been so focused on curing disease that it had given short shrift to prevention. FHS changed all that. Today, we have FHS to thank for making it common knowledge that smoking, high cholesterol, high blood pressure, and obesity are all risk factors for heart disease—along with the lifestyle factors that, in turn, affect these risk factors.

Nicholas Christakis, the Sterling Professor of Social and Natural Science at Yale University, holds an MD and also a PhD in Sociology. This dual training gave him a unique perspective on the FHS data. Because the researchers did not want to lose track of the respondents if they moved, they also asked them for the names of friends who might know how to reach them when the next wave of data collection came around. From his social science training, Christakis knew that many behaviors flow through social networks as if they were electricity on a grid. Why not important health-related behaviors, too? The name data enabled Nicholas Christakis and James Fowler to map a web of relationships for more than twelve thousand FHS participants alongside their BMIs.[7] (BMI is measured as weight in kilograms divided by height in centimeters squared. It provides a useful, if somewhat flawed, measure of health risks for cardiovascular disease, diabetes, sleep apnea, and joint disease—to name a few conditions associated with it. Waist-to-hip ratio and body fat percentage are better measures, but since BMI is so easily calculated, it is more frequently deployed. A BMI over twenty-five is considered overweight and over thirty is labeled obese.)

What they found was fascinating: individuals whose friends were overweight or obese were 57 percent more likely to become overweight themselves. This effect of an obese person's weight on a social contact's weight was much bigger if the contact considered the obese person a friend, as opposed to merely an acquaintance. Moreover, if I consider you a friend, you affect me much more than if you consider me a friend, but I don't name you as one. That asymmetry makes sense, since if I care enough about you to list you as a friend, I am probably more affected by you than if you listed me as a friend, but I failed to nominate you. As most of us may be painfully aware, friendship is not always symmetric. Christakis and Fowler interpreted their findings about friend similarity to mean that, like a flu virus, obesity could be "caught." They didn't think that obesity was literally caused by a bacterium or viral particle in the air like Covid-19. They meant that behaviors like eating habits, exercise, and lifestyle more generally spread through friendship ties the way I might forward my friends a funny cat pic. Perhaps the mechanism was direct: if we are good friends, we might frequently dine at the same all-you-can-eat joint that you recommended when we first met. Or it could be more subtle than loaded nachos every Wednesday night. It could be that by having friends who put on a few pounds, I don't feel so bad when I have to punch a new hole in my own belt to accommodate extra inches.

The resulting 2007 *New England Journal of Medicine* article garnered a front-page article in the *New York Times*. A *USA Today* headline proclaimed, "Obesity Is Contagious." The public debate it launched captured the attention of the U.S. for a brief while. It seemed to explain what any casual observer who had been alive for the past few decades would have witnessed—namely, that all of America was putting on weight. If obesity spreads like the flu, that would explain a lot. It was spreading through communities just like any other disease. Some people wondered if we could actually blame our friends for making us overweight.

The original article has been cited over seven thousand times as of this writing, making it a blockbuster of social science. But Christakis and Fowler weren't the only ones conducting follow-up studies. Economists had recently begun studying the role of social networks as well, usually with respect to job searches, but extending to other domains such as marriage, innovation, information flow, and health.

In this vein, economists Ethan Cohen-Cole and Jason Fletcher were suspicious of the obesity findings, so they decided, as a thought experiment, to model the social contagion of acne, headaches, and height. The pair of researchers showed that according to the original Christakis and Fowler methods, all three of these outcomes appeared to be contagious, just like obesity.[8]

The idea that acne is contagious is pretty suspect. Unless we are rubbing faces together or mimicking each other's lifestyles that lead to a greater or lesser number of pimples, it is hard to imagine how acne would be a socially contagious disease. A tendency toward headaches also seems like a stretch in terms of social contagion. Perhaps one friend gets the other to drink lots of red wine and that triggers migraines? Or maybe the grumpiness of Person A who suffers from chronic tension headaches induces Person B's own stress-related pain? Possible, in theory, but doubtful. While a given headache might be due to the social transmission of a virus, chronic headaches seem more related to individual, idiosyncratic factors, not a friend's headaches.

And that leaves height, which is certainly not socially contagious, as much as some of us may wish it were. Height is 80 to 90 percent heritable. Any environmental influences probably have a lot more to do with prenatal environment, early childhood nutrition, and diseases or medications that might stunt our growth. And moreover, if measured in adulthood after we stop growing, there is absolutely no way the similarity in friends' heights can be due to contagion (at least by friends we met since we reached our adult stature). So, what could be going on here?

HERE'S AN ASSIGNMENT: GO SIT ON A PARK BENCH AND WATCH PEOPLE. In particular, note pairs or groups of people who seem to be strolling together. I like to do this. It's pretty amazing to notice the similarities between people and their sauntering partners. Tall people walk together. Blond people, even. A person with ear, tongue, or nose piercings is unlikely to be chatting with someone dressed in L.L.Bean preppy garb. After a half-hour of observation, I think you will become convinced, like I am, that social life is one big sorting game and that people tend to become friends with people who are like them on any number of observable dimensions—tattoos, weight, hair style, type of dog they are walking, or anything else you can think of with which to classify people.

There's a technical term for all this social sorting: *homophily.* Like likes like. Homophily explains not only why you'll see people matched on looks and style in the park but also why spouses tend to resemble each other in their heights, education, political views, and a host of dimensions that you may or may not be able to observe from your vantage point on the park bench. Homophily is also why groups of teens with shared characteristics sit together in the proverbial lunchroom cafeteria: the jocks, the goths, the nerds, and so on. And homophily also comes into play when small-town residents tend to vote similarly, as do big-city dwellers. Social sorting—like connecting with like—is happening in all domains of social life from our families up to society at large.

The park-bench observations offer an alternative explanation to the patterns Christakis and Fowler observed in their data. Sure, one member of a pair of overweight strollers might have caused the other to become heavy, as Christakis and Fowler had claimed. But it is more probable that people who already are overweight tend to become friends with one another. Likewise, it's likely that homophily, not contagion, explains the lion's share of the similarity between friends in terms of acne, headaches, and heights. People who suffer from adult acne end up being friends by choice or by the fact that they are stigmatized among clear-skinned people and end up together by default. People who suffer from headaches might enjoy the sympathy of fellow migraine suffers. And tall people might find themselves together based on their mutual membership on the basketball or volleyball team, for instance.

If one person's height (or acne or headaches) can't affect the other person's height, then if we happen to see that heights seem to be similar between people who are friends, that likeness must either be because they shared a similar environment (unlikely in the case of height where the environment plays a minor role) or because of homophily. Common environmental influences that happened to be shared by friends or peers tend to be minor on physical outcomes, so social sorting is what drives similarity in most physical traits within social networks.

People who are super athletic tend to befriend other sporty folks. But their homophily could be the result of less direct processes as well. Maybe childhood stress causes binge eating, and friends bond on their shared childhood experiences, which might, in turn, mean that both are more likely to be overweight. The key point is that homophily explains the lion's share of what Christakis and Fowler found, not the influence of friends. Obesity is not like the flu, it's more like a high school cafeteria: people sit with people who look like themselves.

This sort of social segregation is not just about obesity and other physical attributes. We sort based on all sorts of genetic signatures. Moreover, it is happening not just in our immediate social networks but at larger scales as well. Indeed, our whole society is sorted this way. We genetically sort ourselves by neighborhood, state, and region. For instance, some PGIs cluster by state, meaning people with high PGIs for a given trait tend to find themselves living in the same state, rather than spread randomly throughout the U.S. The PGIs for smoking, height, and education tend to cluster the most.[9] Smoking makes sense: people who have a predilection for nicotine may want to live in states where there are fewer smoking restrictions, so they find themselves consciously or unconsciously moving there (or having parents who moved there). Education also makes sense. We have heard from pundits and politicians that we have become two economies in America. The one for college graduates is booming in cities like Austin, Texas, San Francisco, and in other coastal and Sunbelt cities. The other economy is a rural and small-city manufacturing economy that is dying; this second economy is predominantly populated by folks who don't have as much education. Meanwhile, I can only speculate as to why height clusters by state—perhaps it reflects ethnic differences in who settled in which state.

School, friend, marital, and even geographic sorting don't happen through some mysterious process; it's mostly genes hitchhiking a ride as we select where to live or whom to befriend based on observable characteristics. First, we segregate into jobs, neighborhoods, and even regions based on traits like physical appearance, lifestyle, and economics and thus encounter people who are already tilted toward us genetically. Second, even within our microsocial landscape, we tend to be attracted to people who share our traits. Those traits, in turn, are influenced by genes.

However genetic sorting arises, the process blurs the line between nature and nurture, bending and twisting genes and environment into that Möbius strip. Based on our own genes, we actively choose the important people in our social world, thus shaping our social environment. Based on their genes, our interlocutors choose us. Think about that the next time you sit on a park bench.

WHILE WE MAY BE CONFIDENT THAT OUR HEIGHT (OR ACNE) ISN'T affected by the company we keep, for other outcomes, it's not so easy to rule out contagion and implicate homophily (or vice versa). How sure can we be that our friends don't sway us politically or that who we know growing up doesn't affect our cognitive skills or that classmates' tobacco use doesn't influence our decision on whether to smoke? So, when we ask how much of friend or peer resemblance in the world is due to social sorting (i.e., homophily), we hit a problem: If we observe that a married couple or two friends are both diehard Republicans or both extraverted or both drink heavily, how do we know whether they hooked up because of similarity in those traits or, alternatively, whether they mind-melded over years of close interaction?

This may sound like a hopeless scientific dilemma, but we can actually use genetics to determine how much homophily there is in the world. Since genes don't change, we can be sure that any genetic similarity between friends is the work of sorting, not contagion. In this way, we can measure how genetically similar overall friends or lovers tend to be for any trait for which we can calculate a PGI. Moreover, we now have an "X-ray" technology that allows us to see how biologically segregated we are across any scale of society—families, friend networks, schools—even states or regions. When we want to ask why blue states and red states tend to have different political views or why some groups of adolescents smoke and others don't or why spouses seem to be either both happy or both depressed, we can use PGIs as a way to distinguish sorting from mutual social influence.

Using the data from the Framingham Heart Study again, Christakis and Fowler set out to measure how genetically similar friends are compared to randomly paired individuals. They found that we tend to be the genetic equivalent of fourth cousins with our friends.[10] (They factored out race by considering, yet again, only non-Hispanic whites, so racial segregation is not an explanation.) This could mean that—whether we know it or not—we end up being friends with people who are actually our long-lost cousins. The more likely dynamic, however, is that we choose to befriend people who share a number of our traits—like those you might observe from your park-bench perch—and that selection, in turn, filters on the genes of our friends, generating the genetic similarity the researchers observed.[11]

This was a fascinating discovery. Previously, we thought of

genetic relatedness as something about shared bloodlines and common ancestors. If you weren't someone's distant cousin, then you were not genetically similar to them. But social sorting, it turns out, has produced a situation in which we end up being paired with people who might as well be our distant cousins based on mutual attraction. The endless social choice of modern society has ended up reproducing friend networks that are probably not all that different from those in a medieval village from the perspective of the genetic similarity of its members.

Building on the work of Christakis and Fowler, Benjamin Domingue led a study (on which I also worked) that found that U.S. adolescents who were close friends were as genetically similar to each other as second cousins.[12] We dug into the sources of that similarity and found that one-third of the genetic similarity among friends was due to the fact that people who are more genetically alike find themselves in school together. The remaining two-thirds was due to their choice of friends within their school. That is, when parents moved to specific school districts or enrolled their children in particular schools, they ended up sorting them genetically so that teenagers going to a given school were more genetically similar. But the sorting didn't stop there. Once in the school, adolescents further sorted themselves on a genetic level in their friend choices. This finding showed that the genetic similarity between friends that Christakis and Fowler observed was not limited to adults, or even to a single generation. The sorting of parents affected the genetic similarity of kids. In turn, the kids' own choices probably influenced who they ended up staying friends with as adults. Genetic sorting is a multigenerational affair that takes place across all stages of life.

It's one thing to say that teenaged friends are as genetically similar as second cousins, but the causes and consequences of that sorting are not altogether clear. To get a better sense of on what dimensions friends were choosing each other, we wanted to look at sorting on specific genetic signatures. When we looked at friend similarity in the PGIs for height, weight, and education, the results were even more interesting. Of the three traits we studied, friends were the least similar in height—a finding I could have attested to given that I am five-foot-ten and all my friends in high school were six-foot-four or taller. And, likewise, with respect to the genetic signature for height, there was no similarity at all. We could find no pattern of sorting by the genetics of height by school (i.e., height PGIs did not cluster by school) nor sorting within school (i.e., who picks whom to befriend).

But for the genetics of BMI, the story was quite different. As in the case of height, schools did not play a role in segregating individuals by their actual BMIs or their polygenic indices for BMI. But within schools, a whole lot of sorting was going on. We found that friends are almost as genetically alike as first cousins when it comes to the genes influencing weight. The genetics of education told yet another story: overall, friends were as genetically alike as first cousins (who share 12.5 percent of their genes on average). But half of this similarity was due to segregation between schools: American schools are sorted on the genetics of education, whether that's due to admissions criteria or residential segregation on social class background that influences the composition of neighborhood schools. (This analysis, too, was conducted only on white non-Hispanics, so racial segregation was already factored out.) The remaining half of the genetic similarity between friends was due to sorting within a given school. Evidently, the scene of the jocks sitting with other jocks and the nerds having their own table in the school cafeteria is observed not only in Hollywood movies, but also on a molecular level.[13]

The PGI, with its X-ray powers, has revealed the hidden logic of social life. Prior to deploying the PGI, we could be fairly certain only a few outcomes—height, eye color, and skin tone, perhaps—were subject just to sorting and not mutual influence as well. But thanks to the unchanging nature of genes, we can now see the skeleton of social life, the hard bones that are not bent by peer influence but that crisscross and join to form the basis of social structure. The social genome is constantly under construction, and we can finally watch that process.

Our genes select our friends; their genes select us. This may make the social world seem inordinately more complicated than it was before the advent of sociogenomics. However, in some ways, the opposite is true. Namely, now that we have a tool to map genetic sorting, we can take it into account when we look for social genetic effects of mutual influence. We can even do that in the most intimate of settings: the marital bedroom.

WHILE GEOGRAPHIC AND FRIEND-BASED GENETIC SEGREGATION HAS important social consequences, arguably the most important domain of genetic sorting occurs with respect to mate selection. For all human history, people have been reproducing with people who are similar to them genetically. Historically *endogamy*—or the practice of marrying within a specific group—has been the result of the clannish nature of premodern society and the desire to keep inheritances within "the family." The result has been a significant degree of human inbreeding. We may not be like some plants that can self-pollinate, but that hasn't stopped us from finding mates who share our genes.

The Hapsburg Dynasty stretched from Germany in the thirteenth century to the entire Holy Roman Empire in the nineteenth century, making it the second-longest ruling family in European history, after the House of Osman (the Ottomans). The Hapsburgs, like most dynastic families, held onto their power through land purchases, military conquest, and smart geopolitics. But what set the Hapsburgs apart, and made their six-century hold on power possible, is that they married their relatives to a degree unmatched by the other royal families of Europe. Eighty percent of marriages among the Spanish branch of the family were among close relatives—often first cousins or even uncle-niece relations. The head spins at trying to figure out the kin relationships among these royals. The family tree looks more like a tangled bush.

One measure that captures the extent of genetic relatedness within a family tree is called the *F-statistic* or *inbreeding coefficient*. The average inbreeding coefficient for the Hapsburgs was 0.0625; such a coefficient would be obtained if two otherwise non-inbred first cousins had a child. One need not be a child of first cousins to attain that F-statistic, however. Even if one's parents were not cousins, but they themselves were the product of generations of inbreeding, then they would produce a child with a similar inbreeding coefficient. This is how, over time, the F-statistic in the Spanish Hapsburg line increased from 0.025 in Philip I—the founder of the Spanish branch of the dynasty—to 0.254 in his great-great-great-grandson, Charles II. Charles II's inbreeding coefficient is the equivalent of a child born of a marriage between a brother and a sister, even though his parents were "merely" first cousins. His grandmother and his aunt were the same person, for instance.

These marriage practices may have provided a durable hold on power, but they came with severe costs. Charles II's lower jaw was so protruded that his upper and lower teeth didn't even touch. This famous "Hapsburg jaw," shared by many of Charles' relatives, not only impacted their ability to eat but also to talk—since their tongues were typically swollen as well. He had epilepsy and a host of other problems that led to his physical and mental deterioration. When he died at age thirty-eight, childless, the last of the Spanish Hapsburg Kings, the autopsy revealed (with hyperbole, no doubt), that his body "did not contain a single drop of blood; his heart was the size of a peppercorn; his lungs corroded; his intestines rotten and gangrenous; he had a single testicle, black as coal, and his head was full of water."[14]

The Hapsburgs' marriage patterns are only the most extreme and famous example of a practice that has been common throughout history in order to maintain wealth within families. British royals in the House of Hanover suffered from hemophilia. And many less-notable families have nontrivial degrees of inbreeding. Indeed, people have been mating with "their own kind" for as long as humans have walked the Earth. Mostly, that *homogamy* (the technical term for like marrying like) occurred with respect to clan or tribe. The gene pools in which we swam looking for a mate were relatively small, constrained by village life and limited transportation. Even if a community member didn't marry their first cousin, the fact that the community had been mating within a limited pool for generations meant that, from a genetic point of view, the individuals might be more similar genetically even if they were not formal relations. Whatever the cause of genetic similarity, it is risky: most deleterious mutations are recessive, meaning that if you have one copy, they don't do much harm, but if you inherit two copies, then you suffer problems. When you are marrying within the same gene pool, it's more likely that you will mate with someone who shares your mutations, leading to an unfortunate match.

Then came the Industrial Revolution and its accompanying urbanization and modern transportation systems. Suddenly people from once-separated mating ponds met and mated, stirring up the genetic pot. The incest taboo spread in most societies beyond prohibition of first-degree relative (i.e., parent-child and sibling) marriage to frown upon other close kin wedding—uncle-niece or first cousin pairings. The decline of arranged marriages—intended to preserve land and wealth within a clan—also played a role. Lastly, stigmas against marrying outside your religion, ethnicity, or race have also waned. As a result, the frequency of hereditary diseases has declined among most populations in the world.

Today we may laugh at those big-chinned, hemophiliac royals or peasants of yesteryear who married their cousins, but we have almost come full circle. Social norms around marriage and reproduction may have drastically changed since the time when the Hapsburgs ruled much of Europe, but for the average person, the degree of genetic similarity between spouses has not changed all that much—excluding the royal families, of course.

Research led by Stanford professor Benjamin Domingue and others (including me) found that spouses were even more genetically similar than friends—slightly more than second cousins, on average (among white, non-Hispanics in the U.S., at least).[15] Accounting for where people were born—that is, the geographic, genetic sorting of their parents—explains a quarter of spousal similarity in genes. But the rest is who is picking whom. This doesn't mean that your hubby is literally your second cousin but merely that in terms of genetic similarity, they might as well be. In a huge, diverse country like ours, marrying your "second cousin" is pretty darn unlikely to happen by accident. It's the result of assortative mating: the marital equivalent to homophily in friendships.

As remarkable as being married to your second cousin unbeknownst to you is, when we examine the genetic signatures for specific traits—that is, PGIs—we find that we are genetically even more similar to our spouses.[16] On the genetics of height, spouses are somewhere in between half-siblings and full siblings when viewed through the similarity of their PGIs.[17] Meanwhile, for BMI and depression, spouses were no more alike—genetically—than random strangers. So, if a married couple both put on weight or remain thin, it's more likely that such similarities are due to common environments they share or mutual influence on each other. And if two spouses are both depressed, it is again not because people who are genetically more prone to depression (or happiness) find each other in the marriage market but rather because similar ups and downs in life are driving their shared mood, or that one is dragging the other down with them. This was surprising to me: we sort on the genetics of body mass for our high school friends, but when it comes to picking our partner for life and what genes we will pass on to our children, we don't select on those same genes to the same degree.[18]

We like to think that in the modern age, marriage doesn't have anything to do with genetics—that it's a purely social process— especially in the U.S., where the image of the "melting pot" is central to our national ideals. One might be tempted to conclude, then, that genetic inbreeding is a relic of the past—a time of small villages and pastoral living, an era of hemophiliac royals and of rigid social boundaries. But when we look at so-called modern society using our social X-ray technology—the PGI—we find that people are still self-segregating, genetically.

As traditional boundaries of ethnicity, religion, place, and class have waned in the marriage market, and as dating apps have opened up an almost infinite pool of potential suitors, the irony is that, through choice rather than constraint, we have ended up sorting ourselves genetically. We sort on phenotypes—height, education, personality type, and so on—in a very large pool of potential mates, but in that process of matching, we end up with romantic partners who are similar to us genetically, too. The genetic similarities between spouses in the contemporary U.S. are smaller than those in rural Pakistan where first-cousin marriage is still widely practiced, but they are significant, nonetheless. They affect not only who we end up interacting with on a daily basis but also our genetic health as a population. They raise the chances of genetically inherited diseases being passed on. They narrow our genetic diversity, leaving us more susceptible to plagues and the like. And they mean that we are surrounded by people who are not only socially similar to us, but genetically as well—that is, the echo chambers we seem to inhabit end up affecting the very DNA we inherit in our cells.

Another reason for our blindness to our genetic assortative mating is that we assumed the things we often prioritize in mating, like education, politics, and religion, are purely "social" and don't have much to do with genetics. But here the power of the PGI also deflates our vanity. With it, we can see that these dimensions of human existence do have a genetic basis, and that when we sort based on them, we're therefore sorting on genetics. We may no longer marry within clan or ethnicity to the same degree, but we increasingly stratify ourselves on what sociologists call *acquired traits*—education, political views, religion, occupation. And when those traits are polygenic, the result is that we marry people genetically more similar to us than if we were randomly matched up. Simply put, this assortative mating on genetics isn't happening because people are giving each other DNA tests before going on dates (at least not yet); it occurs because the traits on which we sort have a partly genetic basis, so the genes come along for a ride when we choose based on those characteristics.

Social processes as intimate as marriage or friendship selection, then, generate genetic ripples outward to the larger society by sorting the genetic landscape into fractal-like patterns of social structure. These social structures then form the social environment that echoes back on us, influencing who we become, how our own genes play out in the wider world, and, ultimately, what kind of genes and environments we will pass on to the next generation.

IN 1575, THE FIRST COAL MINE IN THE UK OPENED FOR BUSINESS. COAL, relatively abundant near the surface of the ground in Britain, fueled the Industrial Revolution—and by extension, Britain's might in the world. Over the next few centuries coal kept homes warm and powered the factories that were springing up around the country. Men flocked to coal mines to make a living. The work was tough, of course, but it offered opportunity to those who didn't own land of their own to farm. One can speculate that the folks who moved to work the mines were either the most intrepid people of the seventeenth century or the most desperate. Or, perhaps, both.

Fast forward a few hundred years, and those same boomtowns have gone bust as the coal industry withered thanks to the availability of cleaner, more efficient fuel sources. People who could leave for greener pastures have left, contributing to a downward economic spiral that should be familiar to U.S. readers with respect to the Rust Belt and our own coal-mining regions in Appalachia.

When we investigate this social history of migration through a genetic lens, we can see that who left and who stayed in these declining coal mining regions is not random. It turns out that people with high PGIs for education were more likely to leave those dying towns to seek out schooling and other opportunities, while people with lower PGIs for education were more likely to stay put. Then, when those stayers have kids, those kids tend to share their lower education PGIs. Moreover, among the movers, those with higher education polygenic indices tend to move to more economically dynamic locations.

This *economic* boom-and-bust cycle, in other words, has resulted in *genetic* sorting—just one of several ways genes and the environment are two "sides" of the same Möbius strip. Namely, in the UK, depressed coal-mining regions tend to have residents with lower polygenic indices for educational attainment than do more economically dynamic areas of Great Britain.[19] That, in turn, leads to less economic activity in those places. Rinse and repeat.

Meanwhile, a study by Zoya Gubernskaya and myself finds that white (non-Hispanic) immigrants to the United States have higher polygenic indices for education than their native-born white counterparts.[20] It has long been known that even when they are worse off economically, immigrants display better health than native-born Americans on average. One theory has it that the sedentary lifestyle and diet of processed food here in the U.S. explains the paradox. While that may be partially true, we found that the genetic predispositions of immigrants—that they tend to be more genetically advantaged with respect to education which, in turn, strongly predicts health—also explain their hale status. Finally, a study of New Zealanders found that those with higher-than-average education polygenic indices were more likely to emigrate to other countries in search of opportunities.[21]

The PGI for education is designed to predict how far someone goes in school. However, in predicting that outcome, it's picking up a lot of other traits along the way: cognitive ability, of course, but also grit, savviness, ambition, and so on. So, it's perhaps unsurprising that it predicts migration. Leaving everything behind in search of a better life takes a lot of skill, guts, and energy—the same traits (and, by extension, genes) that lead to going far in school. The New Zealand study found that those who moved to Asia were the most positively selected of all emigrees. Moving to Australia, the U.S., or Europe from New Zealand takes initiative. But moving to a country that doesn't even use the same alphabet takes a lot more moxie.

What I've been describing has been labeled the "brain drain" wherein the best and the brightest of a country, region, or town leave for opportunities, removing the most dynamic and talented people from the local economy. This bifurcation of communities is both the cause and effect of genetic sorting. Genetic sorting *causes* polarization because our genes play into where we end up living. But genetic sorting is also an *effect* of the increasingly unequal set of opportunities offered by the thriving locations since that inequality is what induces the gene drain in the first place. In this way, it's a vicious cycle.

Many people are aware of a dramatic increase in overall economic inequality in the United States (and within most countries) over the past few decades. They may be less aware of the degree to which we have also become increasingly segregated economically. For instance, in 2023, the median household income in Washington, DC, was $111,000. The corresponding figure for Mississippi was less than half that, $55,060. A pretty big difference. But even locally, spatial differences are stark. New York City's fifteenth congressional district, which covers the South Bronx, is the *poorest* in the nation, with a median income of $31,000. Meanwhile, the nation's *richest* district, New York's twelfth, is just a mile or so to the south (it includes Manhattan's Upper East Side). Economic sorting and segregation occur even in areas where people of multiple social classes overlap: people of different incomes often frequent different establishments on the same city block, literally walking past each other without interacting. Indeed, one study finds that there is more cross-class interaction in a small-town diner than there is in a diverse, large city with all income strata represented.[22]

What I had no idea about, until I dove headfirst into genetics, was that something else was happening over this period of rising inequality and economic segregation. Not only was social and economic inequality on the rise, but *genetic* inequality was also growing—it was just that nobody could see it back then. It only became apparent after the genomics revolution when we became flooded with data about people's economic *and* genotypic statuses.

There are real, average genetic differences between the coal mining town residents and the residents of London. But what makes them genetically *unequal* is that fact that those differences in DNA affect (and reflect) social opportunities made available under the current organization of British society which gives an income premium to academic skills. It's not that some just have naturally "superior" genes than others across all environments (as someone inclined toward eugenics might conclude). The very fact that certain genotypes have an "advantage" when it comes to education is contingent upon how our society is set up. Moreover, the PGI for education captures all kinds of genes that aren't "rationally" related to cognitive ability, such as genes that influence how attractive we appear and/ or evoke discriminatory responses that, in turn, affect our schooling opportunities. Even the sorting that produces this genetic inequality is also contingent on how society is set up. If great schools and economic opportunities were in the coal mining regions of the UK, then there wouldn't be such drastic geographic sorting or genetic segregation.

Genetic inequality can be spatial, when we geographically sort on our genes as in the case of the coal mining communities versus the big cities in the UK, but it doesn't have to be. By *genetic inequality* I simply mean the distance between the most genetically advantaged people and the least genetically advantaged people for a given trait. So, for BMI we can see that not only is there greater inequality today in how much people weigh—meaning a wider spread in the population's weights—but there's also greater genetically based inequality for weight. In the early twentieth century, when most people got just barely enough calories and a sedentary lifestyle wasn't an option, people of all genotypes were relatively thin. The genetic signature that makes up the PGI for BMI didn't really have any oomph back then: since there wasn't a lot of variation in weight, there were no disparate categories of body types for genes to sort us into, and genes didn't do as much in terms of predicting our weight. But after the environment shifted, the distribution of weight got a lot broader, and the distribution of genes sorted us into those far-apart bins; as a result, the PGI for BMI became a stronger predictor of how big we were.

For smoking, there is greater genetic divergence between smokers and non-smokers today than there was in the 1950s. Back in the day, we didn't know that smoking was all that bad, so people who had non-addictive genotypes socially smoked just as people with addictive genotypes did. When the information emerged about the dangers of smoking—in particular the surgeon general's report of 1964—those who could quit did. The remaining pool of smokers is genetically very different than the non-smokers—harder wired to become, and stay, hooked on nicotine. So genetic inequality emerged: whereas before, the PGI for smoking didn't really separate us into distinct camps that experience different health consequences, now it does.[23]

In these situations, once-benign genetic signatures come to sort us into unequal positions thanks to a shift in the environment. It's not that the genetics itself necessarily gets more unequal; it's more that the effects of genetic signatures in dividing us become more pronounced. But it's important to remember that genetically based inequality is nothing natural or inevitable. If we effectively banned smoking, genetic inequality based on the smoking PGI would disappear overnight. Ditto for inequality resulting from the BMI PGI if we drastically changed our food system. When we make people stay in school by fiat, the impact of genes on educational inequality decreases (as it did when the UK lengthened the minimum years of mandatory schooling its citizens were required to complete). Genetic inequality is wholly contingent on the lay of the social landscape.

But in other situations, the very distribution of genes themselves can also widen or shrink over generations. When the impact of genes gets more important thanks to changes in the environment *and* the very distribution of PGIs becomes more unequal, then we get a double whammy of inequality.

One way the distribution of PGIs becomes more unequal is through genetic sorting via marriage—when genetically advantaged people marry other advantaged people in a process called *assortative mating*. Demographer Christine Schwartz has estimated that increasing assortative mating among U.S. spouses in terms of socioeconomic factors can explain 40 percent of the rise in income inequality in recent decades. What she means by this is that today, high-earning women and men pair up (and women work more than they did in the past), so families at the upper end get a double earnings boost, raising their household incomes. Meanwhile, those at the low-wage end of the economy are more likely to end up together, pooling paltry wages. The result is greater inequality than if we married without respect to earning power—as was more common in the past. Genetic assortative mating is part of this story, too: when ambitious women marry similarly striving men, it also means that the genes for socioeconomic success are being paired up, thereby raising not just income inequality but genetic inequality as well. What's more, this genetic inequality ends up *causing* social inequality when those couples have children. The fact that we genetically sort heavily on DNA signatures that affect socioeconomic success—namely, the education PGI—means that not only is genetic segregation affecting our own generation, causing social and natural inequality to align, it is actually widening gaps in the next generation—by polarizing the genetic bases of economic inequality for our children.[24]

We'd like to think that our romances, social networks, and communities are sites of social mixing and mobility—that coming together with others is a democratic leveling process in which all kinds of people mix freely. But genetic sorting reveals that our social lives aren't as free and integrated as we'd like to believe—that the advantaged socialize with the advantaged, and the disadvantaged with the disadvantaged. When we combine what is happening in terms of the genetics of marriage and childbearing with the non-random migration of genes (in their hosts) across cities, states, regions, and countries, it should come as no surprise that we are increasingly segregated and unequal, not just socially, but genetically as well.

To best figure out what's going on in terms of marital sorting and its consequences for the next generation, it helps to have genetic data not just for individuals, but for entire family pedigrees. That way we can see how genetically alike the spouses are, as well as what the consequences of that similarity are for the next generation. The Scandinavian countries—Denmark, Norway, and Sweden—have long collected reams of data on their populations. They track their citizens from cradle to grave, in conjunction with the lifelong social support the state provides. So, it's perhaps no surprise that one of the largest datasets that has genetic information on entire families comes from Norway.

When researchers analyzed these data, they found that not only

have Norwegians been genetically assorting on the PGIs for educational attainment, height, BMI, cognition, chronotype (being an
early bird or night owl), and smoking—but for some of these traits
(education, cognitive ability, and chronotype), the rate of assortment
has been *increasing* over recent generations contrary to what I had
found in the U.S. with older data.[25] Namely, spouses who married
in the 1920s are less genetically similar on these dimensions than
those that paired up in the 1970s, who are, in turn, less alike than
those wedding today. Despite becoming a more mobile and cosmo-
politan society, genetic sorting at the family level—and thus genetic
inequality—has increased.

We don't know why assortment is increasing for these traits, but
we can speculate. It could be due to increased social and genetic segregation, so that mating pools are more homogeneous with respect to
the genetics of education, say. Or it could be a matter of individual
choice—as education and cognitive ability become more and more
important for economic security and success, people are increasingly
selecting mates based on these traits. In today's economy, if you hope
to have kids who enjoy the best chances for success, your success in
the marriage market is just as important as how you fare in the labor
market. The (genetically) rich get richer . . . even in a social democracy like Norway.

The scary scenario is where these microworlds get segregated
to the extent that genetic differences solidify over generations. The
longer and more thoroughly we sort and reproduce with genetically
similar people, the risk increases that we coalesce into tribes or even
new "ethnic" groups, which, in turn, leads to greater social distance
and animosity. Already there are recognizable genetic signatures, as
noted earlier, between subregions in the UK.[26] Social division leads
to genetic cleavage and vice versa.

All this genetic sorting and resultant inequality has been accelerating without people explicitly knowing what DNA cards they

hold. But once we can actually access our own PGIs and those of our potential mates (and potential offspring), this is likely to accelerate. Before long, social inequality will be written into the genomes of the next generation—unless we do something about it. The wall between nature and nurture is a porous one—before long that border will be completely decimated as social position influences genetic stock (of the next generation) which, in turn, will affect social position. That's because those who have the means are always the first to take advantage of new technologies. Even if the one-percenters aren't genetically different from the 99 percent now, their children will be, and their children's children will be even more. Forget legacy admissions: there's a whole new form of unequal inheritance afoot.[27]

The Social Genome

Before leaving for college, I had to fill out a form asking about my roommate preferences. The form inquired as to whether I was a morning or night person (answer: morning) and whether I was a smoker or nonsmoker (non). After checking the boxes, I scribbled in my own priority: I wanted to live with a "party animal." I arrived in California equipped with little more than this hope—no clothes, no bedsheets—because Frontier Airlines not only lost my luggage but also had seemingly gone bankrupt while I was in the air. But after scrounging some sheets from a store on the way to campus, I successfully arrived at my dorm room at Cunningham Hall. There, listening to the tunes of U2, was my assigned roommate: Anthony De Christoforo.

We were not exactly similar. I was a liberal New Yorker. He was the Republican son of a Reagan-appointed judge in California. But he was the party animal I had requested, and we got along well. Tony belonged to a fraternity, Kappa Alpha, and taught me how to drink beer at a frat party. He showed me how to fashion a fake ID I could use at the bars on the south side of campus. That spring, I rushed his fraternity at his urging. (I was not admitted.) His influence on me that year was profound—much more than any influence I might have had on him. Perhaps that was because he was a grade above me, and California was his home turf. Or maybe it was just because I was a lost soul. After he moved into his frat house and I roomed with someone new (a friend I had chosen), the effects of Tony on me faded. I didn't rush any more fraternities. I drank less beer. I went to the racetrack. I found my footing.

At the time, I had no idea that I was participating in an experiment of sorts. Freshman roommate pairings have long interested economists, sociologists, and other scientists; because of their once random nature, they're a good way to examine the effects of eighteen-year-olds on other eighteen-year-olds independent of their established social circles—at least before nonrandom roommate selection became more common thanks to social media. I can't say in the long run if Tony had a residual effect on me, or if I did on him (Did he make me more conservative? Did I make him more liberal?), but study after study has shown that exposure to roommates does have important consequences on student behavior. Whom you end up living with freshman year affects your drinking behavior, your choice of major, your grades, your weight, and your racial attitudes.

Most of the effects are what we might expect: If you get a roommate who is studious, then your own grades improve.[1] If you get one who is a binge drinker, you drink more yourself, and your GPA suffers.[2] And, if your roommate has joined a fraternity, you are more likely to join one (or, in my case, at least try).[3] More notably, getting assigned a roommate of a different race makes one more tolerant and less prejudiced—while you are living together, and even a few years later. Another interesting finding is that, when it comes to the dreaded "Freshman 15," getting assigned an overweight roommate doesn't cause a woman to gain more weight—it causes her to gain less. (Perhaps seeing the overweight roommate serves as a warning sign, leading to healthier eating and exercise habits.[4] Or perhaps any efforts of the roommate to slim down have spillover effects.[5]) Another study found that the social status and sexual experience of the roommate matters: influence on weight was stronger from higher status roommates and those with more sexual experience.[6]

There are perhaps surprises in the details here with respect to which sorts of roommates have the most influence on which outcomes, but the overall picture jibes well with how most of us—and certainly social scientists—view the world: the people we spend time with help shape who we become. Of course, the physical environment impacts us too. Being around greenery and bodies of water makes us happier.[7] Poor air quality leads to higher rates of asthma and lung cancer.[8] But when we talk about environmental influences, or nurture, we mostly mean the social conditions. I was "nurtured" into my brief enthusiasm for fraternities by my roommate, and had I been accepted into Kappa Alpha, my would-be frat bros would have, in turn, shaped who I became. Moreover, much of our individual and collective achievement is a social process, so it's common sense that the company you keep would matter greatly.

With all this in mind, there is another way of looking at what happens to roommates: it's not just the people we encounter that help shape who we become, it's the genes they carry within them. One can reexamine the same questions of peer influence from the nature side, as if you had donned a special pair of X-ray glasses to see the DNA of college students.

That's what a group of scientists did—only with mice instead of humans. Much like college administrators, they randomly assigned lab mice to be roommates—or rather, cage-mates. The idea was to have these little creatures hang out together and track what kind of impact they had on each other. After six weeks, they measured how the mice had changed—not, of course, whether they joined fraternities, but fluctuations in their weight, anxiety levels, immune function, and so on. The researchers measured a total of 117 outcomes, including the biochemistry of the mice brains.

Unlike the sociologists' who studied human roommates, these scientists knew the genetic profiles of the mice they were sticking in their laboratory dorm rooms. They examined what effect a mouse's genes had on its own health and behavior, but they also looked at whether the genes of the mouse's roommate had an effect. Of the 117 outcomes they tracked, a staggering 43—including anxiety, body mass, immune function, and prefrontal cortex chemistry—showed an effect from the cage-mate's genes. For 8 of those 43 outcomes, the effects of the cage-mate's genes were larger than those of a mouse's own genes.

How could this be possible? How could *another* mouse's genes affect a mouse more than its own? Well, imagine one of these scenarios with humans: There is a limited amount of food available, and your roommate is genetically programmed to have such a voracious appetite that he routinely cleans out the fridge, leaving you with nothing. Or perhaps she has sleep apnea and snores all night, keeping you up. Or perhaps they are super aggressive and cause your cortisol levels to repeatedly spike with their behavior. In the case of the mice, a follow-up study revealed that the specific genes that had direct effects on a mouse were not the same genes that mattered for the cage-mate's social effects on that mouse. In other words, genes for metabolism might matter for the mouse's own weight, but genes in the cage-mate for, say, aggression might affect the weight of a mouse more by causing it to be more restless.

Mice are not humans. But what these experiments revealed has been found to also be true for *Homo sapiens*, a highly social species. In the past decade, study after study using PGIs has shown that the genes of others matter greatly to who we become. If genes are like the engine that drives us to act the way they do, then humans are like bumper cars. We don't stay in our neat, environmentally determined lanes. We careen all over the place, bumping into each other—in college dorms, within our families, at school or work, on the street—and knock both ourselves and others onto new paths with each interaction. By measuring the social genome, we can see which bumper cars are knocking into which other bumper cars and with what force.

"HAPPY FAMILIES ARE ALL ALIKE; EACH UNHAPPY FAMILY IS UNHAPPY IN its own way," Leo Tolstoy famously wrote in the first line of *Anna Karenina*. But is this true? Or are there predictable patterns to household unhappiness too? If you conducted a survey of U.S. married couples, you probably wouldn't be surprised to find that some partners tend to be happy together and some tend to be depressive. Rare would be the couple where one partner is depressed and the other is walking on cloud nine. Indeed, when we compile a checklist of depression symptoms, there is a strong concordance between the scores for wives and husbands.[9]

The question is: Why? Obviously, it would take a sociopath to be unaffected by the mood of their spouse. But what are the reasons that partners tend to be happy or unhappy together? Perhaps people with a propensity toward depression or giddiness find and marry each other to begin with—genetic sorting. Or perhaps the partners have shared a significant event—like the loss of a child or a household financial meltdown during the Great Recession—that has affected both their moods. Or maybe one spouse drags down the other, or buoys them.

There's a 1903 doggerel, "The Work and Wages Party," that speaks to this last idea. Published in the British newspaper *The Sunderland Daily Echo and Shipping Gazette*, it gave rise to a well-known adage about marriage. The verse goes like this:

I'm a work and wages party man,
I say that's what I am.

You'll find me true and hearty, man,
For that is what I am.
Now, let's rejoice to end the strife,
With all the kids in clover,
A happy wife, a happy life.[10]

A more gender-neutral evocation of this last line runs, "Happy spouse, happy house." Beyond anecdotal experience, though, is there a way to test or measure this? We cannot randomly assign humans to marry each other the way we can assign freshman roommates or mice in cages.

This is where PGIs come in. Armed with the polygenic index for depression, we can adjudicate on Tolstoy's behalf the different ways a family can be unhappy. We can measure whether partners found each other that way, whether some external force caused them both to become depressed, whether they affect each other's moods equally per the gender-neutral adage, and whether there's any truth to the idea that it's primarily the wife's mood that colors the relationship.

It is the fact that the PGI for depression is fixed at birth—or rather conception—that makes this causal exploration possible. In 2024, my research group leveraged that fact with data from the Health and Retirement Study (HRS), a nationally representative study of older Americans that began in 1992 and has followed individuals and married couples ever since. In 2006 and again in 2010, the surveyors collected saliva from respondents and extracted DNA from it. Since the sample was recruited before the days of legally recognized same-sex marriages, only heterosexual couples are included in the analysis.

If there was no average similarity between the depression PGIs of a husband and wife in the HRS data, we could rule out possibility one: that people end up marrying people who are similar to them on the genetic predisposition to be depressed. Indeed, we found no correlation between the PGIs of spouses.[11] When it came to depression, people appeared to neither follow an "opposites attract" rule nor a "birds-of-a-feather" norm.

Next, we could see if there was a relationship between one spouse's PGI and the other spouse's actual depression level. This was the equivalent of examining if a mouse's genes affect outcomes for its cage-mate. If the husbands of women with a high depressive PGI turned out to be depressed more than the average husband in the population, it was likely the case that the wife made the husband depressed (or happy, as the adage went).[12] To test this, we first predicted husbands' depression symptoms by their wife's depression PGIs. We found that a wife's genes predicted her husband's mood a third as strongly as his own genes did. So far, support for the old adage: Happy wife, happy life.

Then we turned the causal arrow around. We predicted wives' depression scores by their husbands' PGIs. Here, too, a husband's genetic propensity toward depression had a third as big an effect on his wife's mood as her own genetic tendency did. (Had we found that neither PGI predicted the other's depression, we would have had to conclude that it was shared life experiences that made some couples mutually happy and others both down in the dumps.)[13]

The traditional adage, "Happy wife, happy life," turns out to be only half true. At least in heterosexual marriages that were formed in the mid-twentieth century, the mood in a household flowed equally from men to women, as from women to men. A lesson, perhaps, for men everywhere—or, at least for me, in particular. In 2017, I had remarried and was determined to make it work the second time around: maybe I shared more responsibility for the overall emotional climate of the household than I had previously acknowledged during my first marriage.

Before the ability to measure the human social genome, we were at a loss when it came to knowing why some couples were happy together and others depressed. Was it a wife's childhood experiences of neglect that led her to pick a high-strung husband, whose anxiety, in turn, made her depressed? Or was depression a preexisting condition that both spouses brought to a marriage? The possibilities were almost endless. Freudians would have said it all stemmed back to early childhood repression for both spouses. Family systems theorists, meanwhile, would have insisted that spouses' emotional lives are interdependent—without an answer for who causes what.

Now, however, the PGI provides an image of the social genome that operates beneath the surface of our interactions, revealing the causal arrows of all kinds of influences. It shines light into the black box of the family, allowing us to answer questions about the intimate social dynamics of a household that were, until now, the bailiwick of novelists like Tolstoy or speculative psychologists like Carl Jung. We now have a genetic thread to follow to trace out the contours of social influence—both in our families and beyond.

COUNTLESS MOVIES, TELEVISION SHOWS, AND NOVELS DEAL WITH THE drama of peer groups and risky behavior during high school. From *Rebel Without a Cause*, the 1955 James Dean classic, to 2023's adaption of the classic Judy Blume novel, *Are You There God? It's Me, Margaret*, the impact of peer pressure has been a perennial theme for Hollywood. In the 1985 film, *The Breakfast Club*, the brainy kid who must serve detention initially resists others' entreaties to smoke marijuana. By the end of the movie, however, he has given in and partakes of weed. With this act, he integrates himself into the group.

If one piece of health advice trumps all, it's this: Don't smoke. Among all the relatively common behaviors, smoking is the single biggest factor within our control that affects how healthy we will be and how long we will live. That fact alone makes tobacco a huge public health issue. But to the social scientist, smoking is also interesting because it's inherently a social process. Nobody has an inborn urge to pick up a small paper roll of smelly leaves and take a drag. They are either encouraged or discouraged to smoke by others. They are taught how to inhale. Social bonding takes place around giving someone a light, sharing cigarettes, or taking a nicotine break together. How is the social environment so powerful that it influences our behavior around a substance that is at once highly addictive and highly dangerous?

In a canonical 1953 study, "Becoming a Marijuana User," sociologist Howard Becker describes three steps to becoming a recreational pot smoker: First, a person has to learn how to use the drug to produce real effects (i.e., hold the smoke in long enough); second, they have to be able to recognize the effects (i.e., distortion of time, the munchies, other physical sensations) and associate them with the drug; third, the user has to learn to enjoy these recognized sensations. Some individuals may not even get to stage one for fear or lack of curiosity. Even if they do, they may never master the art of toking on a joint or a bong, so they will not receive the necessary chemicals to induce any psychic changes. In these cases, they will dismiss the experience, claiming that marijuana simply doesn't work on them. Even if they master the art of inhaling, some may not recognize its effects. Or they might not associate those effects—such as excessive laughter or hunger—with the drug. Lastly, even if they can identify the subjective feeling of "being high," they might not enjoy it. They may feel like they are losing control, for example, and worry about having a psychotic break. Or they may become paranoid. Given all these necessary steps to becoming a regular user, Becker describes marijuana as an acquired taste, like oysters or a dry martini.

Becker's key observation is that at every step of the way there is a social process involved. In learning how to smoke, one informant told Becker that he "didn't take his eyes off [a more experienced user] for a second, because I wanted to do everything just as he did. I watched how he held it, how he smoked it, and everything. Then when he gave it to me, I just came on cool, as though I knew exactly what the score was." Even recognizing that one is high often involves social interaction: One guy "kept insisting that he wasn't high. So, I had to prove to him that he was." Another respondent described asking, "What's happening?" and his friend said, "You're on pot." The act of defining those sensations as pleasurable also involves more experienced smokers reassuring novice ones that everything is going to be okay, that they are really not flipping out.[14]

Though more addictive, tobacco use shares many of the same social aspects as marijuana. Namely, one must learn how to acquire tobacco (which is illegal for a minor) and how to inhale to get the effect of nicotine. One must also push through the initial unpleasant experience (usually excessive coughing, eye tearing, and perhaps nausea) to achieve what can be subjectively considered a pleasurable experience. These steps typically involve tutelage, reassurance, and encouragement from others.

Who are those others going to be? According to popular culture, adolescence is a time when individuals are particularly prone to peer pressure from friends.[15] Friend groups provide an emotional waystation of sorts for teenagers who are seeking distance from their parents but who are not ready for complete autonomy.[16] It works like this: People sort themselves into groups. The members of these crowds regulate each other's values, style, actions and so on, and the group becomes the primary source of one's identity. Risk-taking behavior— like smoking or drinking—plays a particularly important role, since the high cost of risky behavior demonstrates (and engenders) identification with the group, solidifying the clique.

Close friends, then, are likely to play a large role in determining who smokes in high school and who doesn't. Whether or not they are aware of this research, many a parent has worried whether their adolescent has fallen into the "wrong" crowd. We instinctively think that our children's close friends are exerting a strong influence on them—right when our children are pulling away from us. We can see it in their styles of dress, the way they talk, and countless other forms of social convergence. But when it comes to something as consequential as smoking, if we observe that Johnny and his best friend both smoke, can we conclude that one of them caused the other to become addicted?

Johnny might have first met his closest friend when they were already both smokers; having cut class to light up in the bathroom, they happened to strike up a conversation. Or it could have been a stressful math teacher they shared who drove them both to need a nicotine fix. Or it could have been Johnny who pressured his friend to take up the habit, rather than the reverse. As with the case of spousal influences on depression, it might seem nearly impossible to answer this question without further information beyond the observation that they both smoke. A high school student is not a lab mouse we can randomly assign to interact with some other nicotine-dependent mouse. The wider social world of a teenager with dozens of friends is also more complex than the two-person system of a marriage. The moment Johnny steps out of the house, he is moving through multiple layers of influence, crashing into other human bumper cars in his neighborhood as he walks to the bus stop; on the bus to school, as he sits in the back row with his best friends; in class, when he is placed in a mid-level track; and at lunch, when he and the entire sophomore grade are given free time to interact in the cafeteria. Moreover, these social layers overlap and fold in on each other.

But the social genome, by providing a genetic thread we can follow to trace out the path of causes and effects, can help us answer the question. As with spouses, if students tend to cluster in high smoking-PGI friend cliques, we can conclude that they probably sorted (i.e., became friends) due to their smoking predilection and need not have had any direct social influence on each other's decision to smoke. But if we look at pairs of friends and find that the smoking PGI of one student predicts their friend's smoking behavior over and above the friend's own PGI, then we know that the student was a bad influence on his friend. And if neither person's PGI predicts the other's behavior but they both nonetheless smoke, then perhaps it can all be blamed on the stressful math teacher.

As it turns out, Johnny's mom can't blame Johnny's new, edgy friend for the fact that she found a pack of cigarettes in her son's backpack. But she *can* blame the school—sort of. In 2019, Ramina Sotoudeh, Kathleen Harris, and I examined whether the smoking PGIs of an adolescent's five closest friends predicted that student's smoking behavior.[17] To answer this question, we used the National Longitudinal Study of Adolescent to Adult Health, one of the first U.S. social science datasets that collected genome-wide DNA information. Much to our surprise, there was essentially no effect. This was a shocker not just because all the psychological literature told us close friends—that tight-knit clique—were the primary unit of socialization during the teenage years. It was also surprising because we hadn't even bothered to factor out the fact that smokers are more likely to be friends with people who already smoke. Even adding together the effects of social sorting and social influence, friends didn't seem to matter all that much.

So, we pulled back to the classroom. Maybe looking at just the friend group was too narrow a focus. After all, the kids in your classes are who you spend the most time with during the school day. We checked whether the PGIs of the other kids a student happened to be in classes with predicted their smoking behavior. Also surprising, those kids had even less of an effect: a big fat zero.

So, we pulled back further. We looked at the entire grade within a given school, where the social composition is totally random—not subject to friendship cliques or classroom tracking. Specifically, we tested whether a kid who attends a given high school with grade-mates who have a higher average genetic propensity to smoke is more likely to smoke herself than a kid who attends the same high school in a different cohort where her grade-mates are less genetically wired to take up the habit. While there may be nonrandom differences in smoking PGIs across schools due to social and genetic sorting, within a given school, it is random luck whether this year's freshman class has a higher average PGI than next year's freshman class.[18] We found that, indeed, a higher average gradewide smoking PGI made a given student in that grade more likely to smoke. Their friends didn't matter. The people they sat next to in math class didn't matter. It was the entire junior (or sophomore) class that mattered—contrary to everything we had suspected about peer influence.

Moreover, we found that the effect of one's grade-mates' genetic tendencies is actually greater than one's own genetic tendency to smoke.[19] In other words, the social genetic effects on smoking are larger than the direct effect of the genes in one's own body. Though I should have been prepared for this result from the studies of mice and marriages, it still blew my mind: the genes of others mattered more than my own genes. The social genome can be more powerful than individual biology—more powerful than we had ever anticipated.

There was an interesting final wrinkle in our high school analysis. Having more so-called "bad apples" in one's grade—that is, having more peers who are in the top tenth of the distribution of genetic likelihood to smoke—makes the entire grade more likely to smoke.[20] Perhaps this is because smoking is highly contagious, in the sense that it is a behavior that spreads through a social network rather than becoming identified with small pods of teenagers. Or it could be simply that a few visible early smokers—the bad apples—set the norms for the entire grade.

So, at least for smoking, perhaps the Breakfast Club of small group peer pressure isn't the right framework, but rather . . . LOLCats (i.e., online cat memes). Humans have been followers of fads from

Holland's tulip-mania four centuries ago to Barbiecore in just the last couple years. We may think that fads are an outgrowth of social media, but the history of the last half millennium tells us that maybe it's something more fundamental about human nature. They suggest that our behavior has always been influenced by others around us—and more specifically, by random strangers.

IF I ASK YOU TO FREE ASSOCIATE TO THE WORD *nurture*, ONE OF THE first images that probably comes to mind is parents. In many ways, we rightly think of parental influence as the end-all-be-all of nurture. Parents seem to have almost unlimited power to shape who their children become by directly interacting with them and by controlling the environments to which they are exposed. My own household provides an example. I remarried to a woman from Bosnia and Herzegovina, and she insists that our son be raised fluent in the Bosnian language. She speaks to him only in her native tongue. Other than seeing his maternal grandparents from time to time and the few hours he spends with a Bosnian babysitter, our son has almost no exposure to that language other than when his mother speaks to him. At age five, he is now fluent. All from the nurturance of his mother.

What language children learn is arbitrary—any child raised from birth can learn to speak any of the Earth's seven thousand tongues. But how fast they learn, whether they will acquiesce to a bilingual upbringing, and many other aspects of language acquisition are influenced by their genes. My wife passed half her genes for verbal ability and parental compliance on to our son, setting the stage for his facility with Bosnian. But there are genes operating in her own body that also affect *his* language acquisition. She showed a remarkable degree of discipline in never answering him if he spoke to her in English. I can't be sure, but I'd bet that resolve has to do with her genes, since I see that same trait in other facets of her behavior. Unlike many other first-generation Americans who seek to assimilate, it was an extremely high priority to her that she pass on her native tongue; this, too, probably has some roots in her personality, and thus in her DNA.

Many other aspects of how we raise our children are affected by our DNA. Whether we are chipper or morose. How we model eating and exercise. Whether we enjoy reading and read to our kids often. And so on. This is all to say that even parenting, the most iconic picture of nurture, is also driven by nature. And yet, of all aspects of the social genome, how the genes of our parents affect us is perhaps the most vexing to study. That's because we inherit their DNA, which means that parental genes do double duty for their offspring: directly within their children's bodies and indirectly, via the parents' genetically driven behavior.

A mother might, for instance, have a high genetic propensity for depression. Her kids will get 50 percent of those depression-prone genes right off the bat. But even before they are born, those same genes (plus the 50 percent they don't directly receive) will be shaping the household environment, perhaps making those children even *more* prone to depression. She might, for instance, take antidepressants or self-medicate with alcohol while pregnant. After they are born, those same genes may influence the overall mood of the household. (If her genes influence her husband's level of depression, chances are they affect the level of depression in her children, too, both directly and through his mood.) She might model depression for them by example, being moody and withdrawn, and she may also fail to provide a nurturing environment, weighed down by her own mental problems, so they grow up with anxieties and a feeling that something is missing.

It may seem like I am picking on moms here. However, for reasons biological and social, the genetic nurture effects for mothers tend to be greater than for fathers. Biological, because they start in utero and continue with breastfeeding. Indeed, research led by Marina Picciotto of Yale Psychiatry shows that in utero exposure to nicotine affects how DNA is packaged, which, in turn, affects brain development and can lead to ADHD and other behavioral issues.[21] Dads' smoking can do this, too, through second-hand exposure, but maternal tobacco use has a much bigger, more direct impact. Moms are more important than dads for social reasons, too, because notwithstanding great strides toward gender equality in industrialized countries, mothers remain the primary caregivers for children in most cases. Fathers, too, exhibit genetic nurture, but the effects tend to be smaller.

Figuring all this out isn't easy because of the intertwining of direct, biological inheritance of genes and the indirect effects of those same genes. The ideal experiment would involve taking human eggs and inseminating them with randomly chosen sperm, then growing those embryos in artificial wombs. Once hatched, those babies would be randomly assigned to eager adoptive parents, for whom we had already genotyped and calculated PGIs. We could then study the effects of those adoptive parents' PGIs for depression on their kids to know the pure genetic nurture effects.

We obviously cannot conduct my idealized experiment. But we can disentangle which genes are passed on and which are not if we have DNA samples from both parents and their children. Once we have determined which genes got passed on from each parent to the child, we can then construct a polygenic index for the "nontransmitted" genes they didn't pass on and see how much of an effect that PGI has on the kids' outcomes.[22] To do these sorts of calculations, you need a big sample of not just individuals, but of entire family units. Fortunately, this is where the small, volcanic island known as Iceland comes to the rescue. Iceland had been unsettled by humans until the Norwegian chieftain Ingólfr Arnarson first colonized it in 874 CE. Waves of other Scandinavian settlers followed. Along the way, many picked up Gaelic slaves and wives from the coast of what is now Ireland. This particular migration pattern makes for a unique genetic signature on the X and Y chromosomes, but that is a story for another book. The point is that Icelanders have a keen interest in their genetics, and in 1996, Kári Stefánson embarked on a project to assay the genotypes of the entire population. His company, deCODE, collected data on almost all of the two-hundred thousand residents of the sparsely populated Nordic nation—without any informed consent on the part of subjects and for private profit, it should be noted.

Notwithstanding the questionable provenance of deCODE, the result is a treasure trove for human geneticists. In the mid-2010s, deCODE shared this gold mine of genetic data with Augustine Kong, a statistical geneticist at Oxford University, who was interested in social genetic effects in humans. Kong and his team went to work figuring out which genes each set of parents in the dataset gave to their children and which genes they didn't. They calculated the nontransmitted PGIs for a number of phenotypes of the parents and found that these had big effects on a whole range of outcomes for their kids, from how far they went in school to their cholesterol levels to their adult height and weight.[23] Of all the genetic nurture effects that were tested, the effects on education were the largest. For educational attainment, the effects of parents' nontransmitted genes were 30 percent of the effects of transmitted genes.

Kong and his team published their findings in 2018 in *Science*, one of the top general science journals in the world.[24] It was a blockbuster paper, and perhaps most interesting was the fact that genetic nurture didn't work in a one-to-one fashion. One might think that for, say, a child's body mass index, the parents' genetic signature for BMI is what matters most in terms of genetic nurture. Maybe parents who have huge appetites stock the fridge with more snacks and such, which might have a spillover effect on children's eating habits; or, those with low BMI PGIs model exercise for their kids. After all, the PGI for body mass index is, by design, the best genetic predictor of BMI in my own body, so why shouldn't it be the best predictor for social genetic effects as well?

But it turns out that the single biggest factor for *all* outcomes—not just educational attainment but also BMI and height—was the parents' polygenic index for educational attainment. Your parents' genotypes for education—even those that they do not pass on hereditarily—affect not only your own cognitive development and schooling success, but also your weight, the age at which you have your first child, your overall health, the number of cigarettes per day that you smoke, and even something seemingly as hardwired as your height. Upon reflection it makes sense. The educational PGI is picking up all sorts of traits that lead to success in school. Those include impulse control and cognitive ability. Having a parent with those traits means you are likely to be fed healthier food (because your parent is likely to know more about nutrition), have better eating and exercise habits modeled for you (due to better self-control on the part of your parents), and be able to afford higher quality medical care (because education translates to higher household income). In short, having a genetically "gifted" parent with respect to education is like hitting the environmental jackpot.

Perhaps less surprising is the fact that it is mothers who matter more than fathers. Less surprising, because even in a highly progressive, gender-forward country like Iceland—literally the most gender-egalitarian country in the world—moms still bear the brunt of childcare. Also less surprising because part of this genetic nurture effect may actually be in utero. An educationally advantaged mother may provide a healthier uterine environment by eating better or avoiding tobacco and alcohol. (Indeed, in studies of adoptees, the maternal genetic nurture effects on kids are smaller, which is likely a result of the fact that adoptive moms do not shape the uterine environment of their children.)

What's more, the social genomic effect of a mom's education genes is not totally explained by how much actual schooling a mom received. Namely, even when Kong and his team compared children whose parents had the exact same level of schooling (at least when measured by years of education completed), there was still a significant, if smaller, effect of the PGI for education on offspring outcomes.

There's another wrinkle to genetic nurture: synergy between the child's PGI and their parents'.[25] In 2015, a few colleagues and I conducted a study on the effects of parental education on offspring using a small sample of genotyped individuals who came out of the Framingham Heart Study, that long and illustrious cohort study of health and aging. What we found was that a kid with a high PGI for educational attainment gets an additional boost from having a high-scoring mom. The same was true on the flip side: a kid with a low PGI for educational attainment who has a low PGI parent is in double educational jeopardy.

Let's take two kids who each scored at the seventy-fifth percentile on the educational PGI. One kid has a mom and dad who both scored at the seventy-fifth percentile. The other kid has parents who both have fiftieth percentile scores, but that kid drew four-of-a-kind in the genetic poker game and ended up with the same seventy-fifth percentile score as the first kid. Kong's model would predict that the first kid will, on average, perform better than the second kid, because the former enjoys better effects of genetic nurture from his parents' higher scores. But our analysis found that it was not just a three-plus-three-equals-six situation; rather, it was more like a three-times-three-equals-nine dynamic. There was an extra, multiplicative bonus for having both a high individual PGI and a high maternal PGI.[26]

The discovery of genetic nurture, and social genomic effects more broadly, represents nothing less than a revolution in scientific thinking about nature and nurture. It represents a challenge to behavioralists who seek to minimize the role of genes, since it turns out that not only do our own genes matter, but the genes of our parents matter too. As Kong liked to joke in lectures about his work, "The best way to get into Harvard is to have a mother who went to Harvard." He was not referring to the controversy over legacy admissions, private school privilege, or generational wealth. Tongue in cheek, he meant that a mother's genotype for education had not only a big effect on her own educational attainment but also a substantial effect on her children's.

Genetic nurture also represents a challenge to geneticists. Before Kong's work, most researchers just tried to predict an individual's outcome by that individual's inherited genotype. But a good 30 percent (or more) of the effect of a person's education PGI, it turns out, is due not to a genetic inheritance but rather to the fact that one has parents who probably have similar PGIs and that *their* genes, in turn, exert a sizable effect. Part of nature as it had been measured up until then, was actually the effect of nurture, not genes acting within an individual.[27] Without considering the social genome, both social scientists and geneticists had been describing different parts of the same elephant.

But it's not just a revolution for scientists. It's a revolution for us all. We have been so used to thinking in this binary—genetics as only what happens in our bodies, the environment as anything "out there" that happens to happen to us—that we are all in need of a more flexible model. It took scientists a good ten years after the Human Genome Project to give up the simple "gene for X" framework and recognize that most human characteristics or diseases are influenced by a multitude of genes. Meanwhile, the failure of the Fragile Families Challenge, as well as findings in behavioral genetics, have meant that social scientists aren't entirely sure where to look to find aspects of the environment that reliably predict outcomes. As for the public, explaining that genes operate more like bumper cars than instruction manuals when it comes to a lot of outcomes will upset both blank-slaters and diehard hereditarians. Blank-slaters won't like the fact that even the effects of the environment are partly driven by genes. Hereditarians, on the other hand, won't appreciate that genes aren't deterministic but part of a messy social process. These opposing ideologues have been walking along the same Möbius strip all along.

6

The Nurture We Seek

The *Toxoplasma gondii* parasite needs cats. Many of us may feel like we need cats in our lives, but *T. gondii really* needs cats. The unicellular organism can survive in other creatures such as birds, pigs, sheep, and mice—in fact, estimates are that up to one-third of humans are infected by this pathogen. While some argue the pathogen is linked to dementia, schizophrenia, and even road rage, most scientists think that it's pretty benign in adults—it just chills out until it gets a chance to infect a cat. That's because although it can survive in other animals, *T. gondii* can only sexually reproduce in the gut of a feline. (What's gross to you is, evidently, romantic to others.)

Evolution seems to have engineered an ideal solution to this problem. Namely, *T. gondii* takes over control of mouse brains. Once a mouse is infected by this parasite, it behaves differently. It becomes more adventurous. A risk taker. The Evil Knievel of rodents. That piece of cheese sitting atop some springy metal thing? No problem. Most importantly, the poor, misguided creature no longer fears felines or other predators; some research even suggests that infected mice become attracted to cat odor. (So, if you wondered why the cat you adopted doesn't seem to have scared away the mice infesting your apartment, it might be thanks to *T. gondii*.)

It's not as if the parasite is pressing buttons and driving the mouse around like a robot. It's the immune response to the *T. gondii* cysts in the brain that causes these behavioral changes—primarily through inflammation in the nervous system.[1] The mechanisms are still not totally understood, but the point is that however *T. gondii* works its magic, it manages to get itself into cats' stomachs to consummate.

What makes the case of *T. gondii* interesting is how it works actively to find a vehicle (the mouse) that brings it to an optimal environment. In this case, the environment is not optimal for the mouse—the inside of a cat's stomach—but that's because it's driven by the genes of the parasite, not the creature's own genetic programming.

The case of the fearless mouse may sound like one of those quirky biology stories we sometimes hear about and file away for regaling (or grossing out) guests at a cocktail party. But the *T. gondii* story is also an example of a very common phenomenon, *active gene-environment correlation* (*active GE* or *GE*, for short). *Correlation* means that two factors tend to go together—when one goes up, the other goes up; when one goes down, so does the other. By *active*, I mean that the genes and the environment are linked due to the choices and actions of the parasite, mouse, or person themselves. Put another way, it is simply the fact that *an individual organism seeks out particular environmental experiences based, in part, on its genetics.*

Examples of active GE correlation abound in the tree of life. Almost any animal that moves to a new environment—or alters its current environment—to improve its chances of survival is causing gene-environment correlation. Every time a migratory bird flies south for the winter, it's engaging in active GE correlation. Even rooted plants engage in genetically motivated selection of their environment when they grow toward the sunlight. Their genes are directing them to change their environmental exposure. Termites build huge mounds to ventilate their underground apartment complexes. Corals make reefs to house their guests—the blue-green algae that use photosynthesis to produce the food that the corals need to survive. Penguins create a warm environment in order to breed in the middle of the Antarctic winter: simply by huddling together, they manage to raise the temperature from negative fifty degrees to thirty-seven degrees centigrade.[2]

For better or for worse, humans, like migratory birds and *T. gondii*, are driven to seek out environments—and maybe even cats—by their genes. Overall, our common human genome makes us explorers, while our genetic differences influence us to seek out distinct environments. In fact, given that we have spread all over the planet and occupy almost every ecosystem imaginable—from the icy Arctic to the Libyan desert—we may vie for the title of most versatile environment adaptor on the planet. Not only have we moved everywhere, we have reshaped the local environments to our liking through processes ranging from the domestication of plants and animals to the invention of air conditioning and the building of massive cities.

In modern society, however, some of those impulses to seek out specific environments can go a little haywire thanks to the mismatch between when those genes first developed (i.e., when we were hunter-gatherers) and how we live now—especially given how rapidly we change the environmental landscape. In fact, one explanation for the obesity epidemic is that most of our evolutionary history took place in a low-calorie environment, so in order to survive, we are hard-wired to be attracted to energy-dense food such as high-fat, high-sugar foodstuffs. But now we live amid a surfeit of calories—sugar and fat, specifically—as well as living more sedentary lives. So those of us who may have been better suited to survival in a desert climate due to our genetic tendency to seek out energy-dense foods like dates and to conserve calories through a slow-moving approach to life are today cursed by that same genetic signature that leads us to Starbucks to order a Venti Caramel Frappuccino with its sixteen grams of fat and seventy-three grams of sugar. Others may find that their own genetically tuned reward circuits drive them to seek out bars or casinos. Still others may be driven to do whatever it takes to get into an elite college.

Anytime we reshape our environs or, more commonly, expose ourselves to a new environment because of our genes, that's an active GE correlation. When you ace your math midterms and decide to push yourself by enrolling in AP calculus and AP statistics in the same year, that's active GE at work—to the extent that performance and drive is a result of your genotype. Likewise, when you perform poorly in summer soccer camp (because of your slow-footed genes) and thus decide not to try out for the high school team that fall, that's active gene-environment correlation because you are choosing to seek environments that do not involve playing competitive soccer. In the case of attending college (or not), your choice is influenced by your traits that have a partially genetic basis—desire, motivation, ability to get in, and so on. In other words, that decision, in part, represents an active gene-environment correlation.

When we step back, we can see that GE correlation further entwines nature and nurture. Not only is there a genetic scaffold undergirding much of the social environment, but also parts of the environment we experience are, in turn, a result of our own genetic makeup. Seemingly straightforward environmental experiences—the cat's gut, AP calculus class, and even who dates us—are not purely environmental at all. Behind the scenes, our genes are guiding us to some environments and not others. Sometimes, our genes shape those very same environments.

In this way, GE correlation reveals something profound: the very way that genes have their effects on us is partly through the environment. There is no doubt that our genes affect whether we experience the environment of AP calculus or that of music lessons, sports camps, and so on. But it's also the case that the genes of people living in the U.S. North are subtly different than the people who live in the South, thanks to generations of selective migration—for all races.

GE correlation paints a fascinating portrait of how humans interact with the world. The lowly earthworm processes the detritus of the forest floor in order to produce topsoil—or what Darwin called "vegetable mould"—in order to create an environment in which it can survive. It's comparatively easy for us to stand above the industrious earthworm and see that it is so much a part of its biome, its physical environment, that for it, genes and environment, nature and nurture, are one, inseparable. It's harder to turn that lens onto ourselves, with our confusing sense of choice and agency. But we are a lot more like Darwin's earthworms than we might suppose. Our genes and environments need and reinforce each other just as much as—if not more than—they do for parasites and earthworms.

OVER THEIR LIFETIMES, BACHELOR'S DEGREE HOLDERS EARN $1,500,000 more, on average, than the typical high school graduate without a bachelor's degree. The average net worth of a college graduate is five times higher than that of the typical high school grad. College graduates even live six to ten years longer than their less-educated counterparts. These are massive differences in life outcomes. Massive enough to make one consider forgoing an immediate paycheck to eat ramen out of a Styrofoam cup for four years to obtain that bachelor's degree—even if one abhorred the vicissitudes of high school.

Alumni of elite colleges fare even better, with early career incomes almost 1.5 times as large as those who graduated from less-elite schools; by mid-career, that premium swells to 1.6. And, as President Biden (one of the few presidents in recent history without an Ivy League degree) told a biographer after his first, failed attempt to reach the White House in 1988, "There's a river of power that flows through this country . . . and that river flows from the Ivy League."[3]

College, then, sounds like a no-brainer, right? It is one of the strongest environmental effects on income in modern society—a key component of nurture when it comes to not just money, but longevity, marital stability, power, autonomy, and any number of other benefits. Being surrounded by smart peers and faculty must surely have a huge impact on the course of your life—not just the skills you acquire there, but the social skills you learn to navigate the world.

But before you quit your job, cram for the SATs, and plunk down a lot of tuition to go back to school, we must consider the fact that it's not random who goes to college. Joe College is more likely than Jack Trade School to come from a wealthier family, grow up in a two-parent household, have more educated parents, and so on. So maybe it is those background factors that make college grads so successful rather than college per se. That is, it does not follow from the preceding statistics that taking someone whose natural inclination is to skip college, plunking them in college, and forcing them to get a degree will result in the same income premium. Or that artificially preventing a college-bound senior from attending university will lower that person's future income.

Even if we compare people within the same family, the differences between those who graduated college and those who didn't are not trivial. Leila, the agoraphobic sister from my old neighborhood, didn't attend college, while her younger sister, Crystal, not only went to an elite college but got a PhD as well. However, it may not have been their educational differences that set their lives on such different tracks. Crystal's PhD could be an *effect* of their (innate) differences, not a *cause*. Ditto for Bill Clinton—a graduate of Georgetown University, Rhodes Scholar, and Yale Law '72—and his ne'er-do-well half-brother, Roger, who never finished college. The point is that Crystal and Bill might have succeeded—compared to their siblings, at least—even if college hadn't been an option.

Perhaps through their rigorous selection processes, colleges pick people already on their way to becoming socioeconomic "winners" and then take credit for them? Or is there actual value added by going to college, in terms of future labor market success? That value added might simply stem from the sheepskin effect, the status marker that is sought after by employers. Or it might also be the actual hard and soft skills one learns at college. Or the value of a peer network. Or having spent four years in the nurturing bosom of campus before having to face the harsh realities of life. How college might have a salutary effect is a second order question, however; first we need to know whether college actually has a treatment effect or whether it merely stamps the passports of the soon-to-be successful on their life journey.

The problem with knowing whether college is really all that it's cracked up to be lies in GE correlation. The value of college seems like a straightforward question about the environment, about the impact of a particular institution in society. However, our innate nature is mixed up in the question, too, by influencing who attends college and who doesn't. We cannot simply compare college graduates and non-college-graduates to nail down the purely environmental effects of college—because our genes are tangled up in that environment through active GE.

How do the genes select the college environment? Some part of it is self-selection: in the way we currently organize our schooling system, those with a high-education PGI are more likely to take an academic track and apply to college. Once they are there, they are more likely to stick to it and complete the degree. We don't know for sure the exact proportions of the education PGI's effects that run through, on the one hand, better problem-solving ability, and on the other, tenacity and charm. We do know that both cognitive and non-cognitive skills are being picked up by the index. Another part of the PGI-college connection happens at the admissions offices, not in the bodies of the aspiring students. Through whatever signals they see in applications, colleges—and particularly selective colleges—end up choosing to admit applicants with higher education PGIs. They somehow recognize genetic potential without the benefit of a DNA swab and grab it through their admissions processes. The written information in the application is, among other things, a sort of abstract nasal swab that allows admissions officers a glimpse at the applicant's DNA.

Measuring people's PGIs lets us address this question from a novel angle. Rather than trying to find a natural experiment (like comparing identical twins who differ in their education levels), we can statistically compare the incomes of high-PGI people who didn't go to college with high-PGI people who did attend college. And ditto for low-PGI people. If attending college gives an income boost even when we're considering a sample of people with the same PGI level, that suggests that what transpires at college is partly responsible for the boost. If there isn't an income boost net of PGI, then it suggests that college is just a place people with higher educational PGIs go to keep busy for a few years. Meanwhile, we can also see how much of the difference in income between high and low PGI folks is contingent on going to college—that is, how much those genes work through the environment of college attendance.[4]

When we do this analysis, we find that if we compare two kids with the same PGI, one who attended college and the other who didn't, there is still a boost for attending. In fact, it's bigger than prior twin-difference-based estimates (where identical twins who complete different levels of schooling are compared in terms of their earnings).[5] If it had turned out that none of the effect went through college, we would see that the PGI would predict the same wages for someone regardless of whether they went to college or not. Moreover, among two people who have the same PGI and both attended college, the one who went to a more elite school will likely have 25 percent higher wages twenty years later.[6] We can run a final check on whether the effect of genes "needs" to run through the environment of an elite college by deploying another PGI: by testing the PGI for income. In this exercise we also see that the genetic path to income mostly runs through college.[7]

In sum, a college degree is not merely a badge awarded by admissions officers for having good DNA; it may not be worth the full 1.5-million-dollar cost, but a good chunk of the income premium for going to college is indeed the consequence of actually attending college. I, frankly, was surprised that college mattered as much as it did in our analysis. As a skeptical professor at an elite institution, I was shocked that attending a highly selective college had such a big effect on my students' wages. Some of my students are brilliant coming in, and some are less so. In some classes they may learn technical skills that employers value—statistical analysis, in my case. But they could have learned these same skills at a local, public college or, better yet, on the job. The stakes of these results are, in my view, high: the endless discussion about college access, and more particularly, about elite college admissions practices, is not just a distraction when it comes to economic opportunity. College really matters.

The even more interesting part of this story, however, is the converse: not just that college adds a boost over and above the effect of the PGI, but also part of the effect of pro-education genes on wages depends on going to college. The genes work by causing their host to select (and be selected by) the environment of college—and better still, an elite college. This is a case where our genes (those for cognitive ability, sure, but also those for attractiveness and any number of other factors that affect our path through the educational system) lead us into an environment that is critical to the realization of the phenotype. It's not that genes matter, *and*, independently, the environment also matters. It's that genes guide us to an environment that they need, in part, to become actualized. When we go to college, we are not all that different from the *T. gondii* parasite; our genes are guiding us to the environment we need for success. We may regard the campus of Princeton University as analogous to the intestines of a cat. The interbraiding of nature and nurture isn't just something that "happens." It is something we plait ourselves: guided silently by our genes, we weave a pathway through the myriad environmental choices and challenges we face, creating active GE correlation as we seemingly meander through life. Our meanderings, when viewed with the PGI X-ray machine, are not so random but follow the same molecular logic that guides the earthworm or parasite.

ASIDE FROM ACTIVE GE CORRELATION, OUR GENES END UP SHAPING THE nurture we receive in other ways. The first is called *passive gene-environment (GE) correlation.* A passive GE correlation happens when your ancestors' genes—often those of your parents—affected the environment for you. Unlike active GE, passive gene-environment correlation occurs not because of that organism doing anything itself, but rather because some of its genes—when they were in the bodies of its parents or other ancestors—caused those progenitors to end up in environments that affected their descendants. The child (or grandchild, or great-grandchild, etc.) is the passive recipient of the choices and constraints faced by its genetic predecessors.

The cuckoo bird, for instance, is a parasitic egg layer. Mother cuckoos choose the nests of other birds—other species altogether—in which to deposit their eggs. The mom essentially gives up her future children for adoption; she may, in fact, spread her eggs across fifty different nests. Some cuckoos specialize in one host species. The egg colors and even how the chicks look once they are born evolve to match those of the host birds in whose nests they will be deposited. So, baby cuckoo chicks hatch to find themselves in an environment that was determined by the genes of their mothers. It was the choices of the mama bird and her genetic predisposition that put them there; that's what makes it passive. But the fact that they inherited half of their mother's genes creates a nonrandom linkage between cuckoo chick genotypes and what kind of bird nests the chicks find themselves in. Their own genes went along for the ride.

Many of the childhood environments we find ourselves in due to historical circumstances, limited opportunity, social inequality, discrimination, our state, our neighborhood, the K–12 schools we attend, and so on result in passive GE. That's because even though genes may not be the cause of these unequal environments, they are not randomly distributed across them. In some ways, passive GE is directly connected to genetic nurture: through the choices genes influence in the bodies of our parents, the genes of our parents shape our environments and affect who we become. But passive GE can extend back for many generations. For instance, if it was our entrepreneurial great-great-grandparents who immigrated to the U.S., then they are exerting passive GE from the grave—their genes drove them to choose an environment for us, four generations later.[8] We might be tempted to call that genetic nurture by ghosts. It may not have been their actions at all, however: it may have been society's reaction to genes for skin tone that confined our grandparents, and then our parents, and then us, to segregated neighborhoods, for instance.

The end result of passive GE is that genes are far from randomly distributed across environments. African Americans in the U.S. North have more European ancestry, on average, than those living in the South. This pattern likely exists because those African Americans who had both white and Black parentage were, in general, more privileged than those with more exclusively Black family histories and thus had the resources to move when the opportunity arose during the Great Migration of the early twentieth century. Furthermore, those who moved North had more opportunities to marry and bear children (voluntarily) with people of European descent once they arrived. The differences in average genotypes today between Black Americans in the North and those residing in the South are not a result of everyone moving themselves in their own lifetimes. They're due to earlier generations of movement. So, the current generation of African Americans who live in different environments (New Jersey vs. Alabama, for example) who track somewhat with different genotypes are experiencing passive GE due to their ancestors' active GE. Meanwhile, if you grew up in Minnesota because your parents avoided sunny climes because they genetically have very fair skin, your state is correlated with your genes through passive GE correlation.

Passive GE correlation can even play a role in environments that we seem to choose actively. If you got into college because of legacy admissions or simply because you grew up in an environment that valued and expected education because your parents (and their parents) went to college, that is also a form of passive GE correlation. The environment—though related to the genes you inherited from your parents—might not even be available to you if those same genes didn't reside in your parents and drive them to matriculate at that college or university. Moreover, passive and active GE are at work simultaneously in this case: we actively choose to study hard and apply to a certain competitive school and that environment has been made available to us through those same genes working in our parents.

In different ways, both active and passive gene-environment correlations show how nature and nurture are not entirely separate. I once did a study and found that whether you were born in a rural or urban environment could be predicted by your genes; how this association between genes and place came about is likely due to multiple factors. Obviously, kids generally don't have any say in where they are born, but migration patterns of generations past could induce the genetic connection through passive GE. It can even be the case that a natural disaster affected a locality's gene pool, and we observe its effects for generations hence. The point I was making in the paper that resulted from the study, however, was that even the most seemingly pure environmental factors, totally outside of our voluntary control, are connected to our genes.

MALE BIRDS ARE PERHAPS THE MOST COLORFUL CLASS OF ANIMALS IN the world (tropical fish give them a run for their money, though). They come in all colors of the rainbow, sometimes all in the same little creature. The genetics of their pigmentation and how the microstructures of their feathers are designed to reflect certain wavelengths of light is just beginning to be understood.[9] But we know for sure that the genotypes of songbirds matter for their feathers' coloration. These physical qualities, in turn, affect how mates and competitors respond to them—the more colorful birds tend to attract more attention from both potential mates and enemies. The environmental response their genes evoke—a mate coming toward them, or a competitor flying away—is related to their genes, but it's not them that are altering the environment directly. It's their genes evoking an environmental response.[10]

In this way, not only do our genes affect our environment through our own actions—they can often evoke an environmental response among others. *Evocative gene-environment (GE) correlation* is when we experience something that results from how others react to our genes—as those genes affect our appearance or behavior. Our genes for height might cause people to choose us first for pick-up basketball. Our genes for beauty may, as with the songbird, inspire many suitors to come knocking. Our genes for extraversion and humor may reward us with many dinner invitations. If in active GE correlation, we choose our environments, then in evocative GE correlation, our environments choose us.

Because a genetically evinced environmental response is evoked in some other organism, evocative GE correlations are inherently *social*. Active GE linkage may act upon nonsocial aspects of the environment—such as a home built to protect one from the elements—but if your genes for halitosis make people not want to hang out with you, an evocative gene-environment relationship develops between the genes for bad breath and social isolation. Ditto for any social response to any trait or characteristic or disease that has a genetic basis.[11] Since humans are highly social animals, this sort of environmentally evocative effect of genes is rife in human populations. Depressed people evoke a response of avoidance (or sympathy) in their social worlds. Funny people evoke laughter and good humor. Tall and short people generate different social reactions as well— taller people are seen as more attractive, competent, and natural leaders. Basically, any trait that has a genetic basis and to which other people react in a somewhat predictable fashion falls into the bucket of evocative gene-environment correlation.

From another perspective, the genes in ourselves that evoke environmental responses in others are the same genes that form part of their social genome. Our genes are the genetic environment they are encountering. In this way, much of human social life is like birdsong or an African American church service: call and response. We navigate through environments in our genetic bumper cars. Our DNA guides us—bouncing off the other bumper cars around us; those cars, in turn, clear some pathways and block others as they respond to us in myriad ways. This social response, of course, is driven in part by the genes of others—the social genome.

In considering the social genome, we saw how others' genes shaped our social environment. Namely, the social genome was the DNA of others affecting us. Evocative GE takes that entwining of nature and nurture one step further. Our own DNA affects others by eliciting a response. That response, which is partly based on their DNA, shapes our social environment. There is a constant tango between our genes and the genes of others.

Long before you submit your college application to the admissions office, your genes have been not only guiding you toward (or away from) certain life paths, like AP classes, but they have also been affecting what paths are open to you by determining how others treat you. Even when it comes to the people who love you most: your parents.

That our genes affect how our parents parent us is a radically new idea, at least among scholars. For the last quarter of the twentieth century onward, the reigning model of child development was the Ecological Systems Theory developed by Urie Bronfenbrenner of Cornell University. According to Bronfenbrenner, the primary caregiver (traditionally assumed to be the mother) was the most important source of nurture for a child. In a matryoshka nesting doll system, the primary caregiver is, in turn, affected by the rest of the nuclear family, whose norms and values are shaped by the extended family. That extended kinship network is, then, affected by the outer social world.[12]

By and large, Bronfenbrenner's model hewed to the blank-slate theory of human development: that children arrived in the laps of their caregivers without any unique prior programming, but only with universal needs, wants, and desires, and hence most differences could be explained by how they were raised.[13]

For decades, there was no real way to test the Bronfenbrenner model of child development against one that suggested a greater, more active role for the child in evoking forms of nurture based on their genotype. Now, however, with polygenic indices, we can answer a host of questions about how children affect their parents' behavior and attitudes. Since we know that parental behavior (or, really anything else about the environment) cannot cause the child's PGI score to change, we know that in any relationship between the child's PGI and the parent's attitudes or behavior, the causal arrow must point from the child to the parent, and not the other way around.

In 2020, my Danish colleague Asta Breinholt and I got ahold of data from the Avon Longitudinal Study of Parents and Children

(ALSPAC), which genotyped the family members of its respondents. We wanted to know if a child's PGI for educational aptitude caused their parents to raise them differently, net of the parents' own PGIs. If so, this would upend the idea of children as blank slates. The experiment works like this: Take two parents who are both at the fiftieth percentile for the education PGI. If they had a hundred children together, the average education PGI for their brood would be at the fiftieth percentile as well. But any given kid can deviate from that mean level. Would they parent a kid who was at the fortieth percentile—by random chance—different than they would parent a child who ended up at the sixtieth percentile?

We examined parenting in early childhood—from ages six months to fifty-seven months. This is thought to be a critical time for child development. Research has shown that early interventions in the environment of children below the age of five yield enormous payoffs in terms of how those kids turn out—whether measured by test scores, graduation rates, criminal activity, or even wages. Moreover, this is a time when there is little institutionalized feedback on a child's progress. There are no report cards or test scores to stigmatize or laud a child, so the parent is responding natively, so to speak, to the informal signals sent by the child.

ALSPAC asked the mother how frequently she engaged with the child in specific activities that are said to be conducive to healthy child development, such as reading stories, singing songs, and playing with the child. As the child grew up and was periodically reassessed, we found that a child who draws a higher-than-expected genetic quotient for educational attainment evinces more "positive" parenting than their sibling who gets a worse hand in the poker draw of human genetics: each additional standard deviation in child PGI increased the amount of time that parents played with their kids by 6 percent—an effect that is about a third as big as the parents' own education levels and which is larger than the educational PGI of either parent. The child's genetics matter more than the parents'.[14] How does a two-year-old communicate its needs and proclivities to a parent? It might be that this is a more demanding child, talking constantly to their parents to draw them into their world. Or it might be that a smarter kid is more interesting to play with—for instance, they might invent more compelling scenarios and fairy tales. Or, they may be more verbally adept.[15]

That children's innate characteristics affect how we parent is somewhat interesting, but probably not all that surprising to anyone who has more than one child and has seen how kids from the same two parents, raised in the same household, can be so, so different. But that the educational potential of our child affects how much we invest in that child from before they can even form complete sentences— now, that's something that many contemporary parents might not feel comfortable admitting. At least, I, as a father of three, do not. Moreover, the fact that we see that our children's genes are teaching us how to parent them from such a young age seemed remarkable to me. As a parent, I had thought I was more in the driver's seat, but upon reflection, I could recognize how differently I raised all three children, thanks to their own demands and needs. From the point of view of the child, this is evocative GE, but from the point of view of the parents, it's a social genomic effect—a reverse generational genetic nurture of sorts—because it's the child's genes affecting parental behavior.

Our study didn't examine how consequential the parenting behaviors we measured were to a child's cognitive development. But to the extent that these behaviors do matter, our results mean that part of how the child's genes lead to their developmental outcomes is through the responses the child evokes in the parents. This is perhaps one of the most important evocative GE correlations in modern society. At a critical age for development, genetic variation in children "tells" their parents to alter each child's environment, which has the potential to snowball over the course of their lives. A kid who evokes cognitively stimulating early childhood nurture may learn to talk sooner or read more quickly. These skills, in turn, may open up other developmental opportunities for them—for instance, to have more sophisticated Socratic interactions with their parents and other caregivers. Those interactions, then, hone their mind even further. It may be that the response a high PGI evokes from a parent is a critical pathway from genes to educational outcomes twenty-odd years later. This possibility isn't something we could test directly in our study, but it is suggested by our understanding of how genes work through the environment to achieve their observed outcomes. When the children we studied grow up, we may be able to tell how crucial the parental responses to their genotypes were to how they turned out, educationally speaking.

Certainly, the benefits of early parental attention have long-lasting consequences that may snowball. But social life is hardly ever static. The same genes that evoke more reading or playing at age two might evince a completely different response when they are teenagers. Indeed, a 2023 study by economists Anna Sanz-de-Galdeano and Anastasia Terskaya found that in adolescence, siblings with *lower* education PGIs get more investment from their parents in terms of time and effort—the reverse of what we found in early childhood.[16] Namely, comparing two siblings from the same family, the one with the lower education PGI is *more* likely to be brought to museums, plays, and other activities; the lower PGI sibling also gets more emotional attention and help with school.[17] Adolescents' genes, then, evoke quite different responses from their parents than they do in early childhood.

During their children's early years of life, it seems that parents are instinctually reacting to the behaviors of their offspring, being led by them, in a sense. By adolescence, with a long paper trail of report cards, test scores, and overt parent-child power struggles, parents may be in a different mode altogether. By then they could

be actively trying to compensate for their children's genetic disadvantages. Any parent of teenagers knows that it's not the A-student
who gets your attention; it's the unmotivated kid who is struggling
to pass their classes. As much as parents may amplify sibling differences in educational trajectories during their early childhood evoked
responses, they appear to be actively trying to undo those differences
a dozen years later. What is key here is the rhythmic dance that genes
perform with the environments around them: there is not "one" relationship between your genes and the environment. It is a wending
path that doubles back on itself, folding into that Möbius strip we call
the nature of nurture.

WE MAY NOT THINK IT IS FAIR OR DESIRABLE FOR PARENTS TO GIVE
more attention to young kids who show more educational potential. But at least it makes some sense. We can consider how parents
respond to their child's DNA draw with reading (or playing) behavior
to be *rational* in several senses. It seems rational on the most basic level
for a parent to respond to a child who wants to hear more books by
fulfilling that desire. It's also rational for the parents to spend more
time with children who are likely to benefit most from that investment. In short, adopting different parenting strategies based on the
child's genetic predisposition can be seen as a logical response to the
child's aptitudes.

Even the peahen who mates with the peacock sporting the most
brilliant fan of tail feathers is rationally reacting to that peacock's
genes. The fecundity of the male's genes is signaled by the brilliant
colors in his plumage, even if those feathers may be of no practical
use and are, literally, a pain in the ass for the peacock. Specifically,
bright colors and long feathers mean that a potential sperm donor
has a well-functioning genome—he is capable of gathering enough
nutrition to produce such a display and of successfully building it

from his DNA—not to mention that he is likely to produce sexually attractive male offspring. It's no surprise that the peahen falls for it. The peacock's tail is a classic example of what ecologists call *costly signaling*—organisms trust signals that are costly to transmit, as is the case for the tail feathers.

However, evocation is not always rational—meaning, it may not be in the best interest of the organism whose genes are doing the evoking in others. Similarly, evocative GE may not even be in the best interest of the person (or other organism) who is being prodded to action by the genes in another; these reactors simply may not be able to suppress a nonconstructive response. Kids who are attractive—as encoded by their genes—get more attention and praise from adults. They are seen to be more honest, trustworthy, intelligent, and sociable. Yet evidence shows that this is a false belief. Ditto for attractive adults. An employer may not be able to resist hiring the most physically attractive applicant even if other evidence suggests they may not be the most productive. When experimenters create fake, online job applications where they vary how someone looks with all other aspects of their resumes (education, experience, skills, and so on) being equal, they find that the attractive people garner higher ratings, more callbacks, and more offers. Since such experiments were controlled, we know that applicants were being judged solely by their looks, not by their actual qualifications for the job—which isn't in anyone's rational interest.

Another classic piece of evidence that shows that some of our judgements represent irrational evocations from people's genes comes from music auditions. During the 1970s, some major orchestras began to conduct "blind auditions" wherein an opaque curtain was strung between the auditioner and the judges. It turned out that many more women were selected under the new regime than in the prior regime. Systematic analysis by economists Claudia Goldin (winner of the 2023 Nobel Memorial Prize in Economic Sciences) and Cecilia Rouse (later President Biden's Chief of the Council of Economic Advisers) showed that orchestras that adopted the new approach increased female membership by 50 percent on average. The proportion of minority musicians also rose. One can easily imagine that not just gender and race but also attractiveness mattered less. As a result of this study, today all major orchestra auditions are performed blind. Who said that academic research doesn't translate into policy? Orchestras have a clearer mission (assemble the best musicians) and better measure of performance (an audition) than college admissions or most jobs do, but this "blind" approach certainly can be and has spread to other domains. NASA now has blinded grant review. Some employers are experimenting with machine learning evaluations of resumes (notwithstanding biases baked into algorithms) and even blinded interviews.

But even if successful, these tweaks won't eliminate all irrational evocation from who wins and who loses in society. That's for two reasons. First, the audition behind the curtain is only the final test after a long process of talent development. By the time someone is playing the cello for a slot in a major orchestra, they have been encouraged, given opportunities and feedback, and have internalized the reactions of others for their entire careers. The evocative response of others—based partially on irrational factors—may have fed back into their confidence and even their musical ability. In the same vein, we are, in fact, admitting students to Princeton based on height, body mass index, beauty, charm, and athletic talent (even for nonathletes)—even without considering the interview portion of the application process. Even if admissions officers are trained to ignore traits such as those, the rest of society has not gotten the memo. Beauty, height, and so on have been evoking responses for the entire lifetime of college hopefuls up to the moment they press the submit button on their application. Beautiful students' grades and test scores are higher, on average, in part because of the positive investments they have received thanks to their pulchritude. Irrational genetic evocation is rife throughout society.

Moreover, in a post-industrial economy, it is not always so clear cut what's rational and what's not. YouTube influencers, for instance, tend to be physically attractive. That's because while they may be conveying information such as product reviews to their followers, they are also watched for their youth and beauty. Is the charming interview with the writer part of the product we buy when we purchase a novel? Hard to say, since we may have never known about the book absent seeing the TV interview.

In this manner, irrational evocation is all around us. So many genetically influenced aspects of how we look and behave evoke responses from the people who make up our social environment. Any aspect of how we interact with or appear to others that is influenced by our genes, but that may not really relate to how we *should* be judged, falls into this category of irrational evocation. Skin tone. Height. Eye color. Voice timbre. And so on. When the wall between nature and nurture falls, you can see that genetic effects don't have to be rational. They can be terrible, too.

In theory, we don't like to judge people for their genetics—something they cannot control—yet such judgments are pervasive in social life. While 92 percent of people tell the General Social Survey that it's unfair to judge people on the basis of genetics, we talk about "natural" gifts or talents all the time in our discourse about athletes, celebrities, politicians, and the like.

For instance, it has long been observed that within the African American community individuals with darker skin are more likely to suffer from high blood pressure (hypertension). This effect most likely results from an increased burden of social discrimination when compared to lighter-skinned individuals. Skin tone, in turn, is controlled by genes. In the study on skin tone and hypertension I conducted with colleagues, we compared siblings, one of whom ended up genetically darker by chance, and found that the darker one had a higher blood pressure, on average.[18] We also checked that the genes for skin tone didn't cause hypertension through direct mechanisms on the cardiovascular system among non-Black people who would not suffer from the social effects of irrational evocation.[19] They did not. So, the genes for darker skin tone "evoke" higher blood pressure via the mechanism of social discrimination. It's weird to think of that as a "genetic" effect, but the syllogism is as follows: genes (mostly) cause skin tone; skin tone provokes discrimination from some interlocutors; discrimination causes stress; and stress causes high blood pressure. This chain of causation is no different than the precocious child who gets put in advanced classes—in both cases, genes evoke a particular response from others. The only difference is that it's rational and at least arguably admirable to put the precocious child into the honors track. It's irrational and immoral to treat someone differently on the basis of skin tone.

Balding men, too, evoke an irrational response from the social environment in the form of the discrimination that they face in the labor market. By some estimates, there is a 5 percent wage penalty for male pattern baldness in early adulthood. Since that's almost entirely genetically determined, it's another evoked genetic response. People who are tall enjoy a wage bonus, just like most groups of thin people do.[20] Managers who judge and reward their employees based on productivity will post better numbers than managers who promote less productive workers who are taller, thinner, and more handsome. And yet . . . it's very hard for all of us to purge biases from our mental frameworks.[21]

Such judgement calls about the irrationality of selection criteria do not need to be limited to genetically influenced factors. Take the example of the "ban the box" movement. Many activists worried about the disparate impact by race of the question about criminal convictions on employment application forms given the higher rates by which African Americans are swept up in the criminal justice system. At the same time, employers rightly would like to make sure the people they hire are not likely to steal from them or otherwise commit crimes while in their employ. It turns out that when the box was banned in certain places, hiring of Black people decreased. Without the explicit piece of information about criminal records, employers hedged their bets through a process called *statistical discrimination*. Knowing that African Americans are more likely to have a criminal record, and with no easy way to find out if a particular individual had a conviction or not, many employers consciously or unconsciously bet on group averages and hired African Americans less often. In this way the policy had the opposite effect as intended, exacerbating rather than ameliorating racial disparities in employment. Statistical discrimination is rampant in society. Every time we make a decision, we are unconsciously calculating group probabilities—there just happen to be so many groups that each person belongs to that we might arrive at a unique judgement for that particular person's likelihood of x.

PGIs can actually help us understand (and get rid of) those irrational biases. When we run a genome-wide association study to construct a PGI for an outcome—hypertension, education, income, you name it—the statistical analysis is agnostic as to whether the effect is rational or irrational in our society. The PGI for income includes any effects on income from the genetics of baldness, height, and BMI; the PGI for hypertension captures skin color (in the Black community); and the PGI for education captures early childhood cuteness—all thanks to our irrational bias to favor people based on these aspects of their looks. That's not the fault of our genes. It's the fault of our unfair society. When we see that the PGI for baldness affects wages for men but not for women (who don't suffer the PGI's effects on their scalp), we can be pretty sure that the PGI has detected appearance-related discrimination.[22] Ditto when the genes for skin tone have an effect on stress and hypertension in African Americans but not in other populations. The PGI X-ray machine can call us out on our collective bullshit.

There are ways to try to suss out whether a PGI's effect is "rational" or whether irrational evocation is slipping in. For instance, a scientist can look and see where the genes that are the most heavily weighted in the PGI tend to be expressed—in which tissues they are the most active. As we would expect, the education PGI tends to be the most heavily skewed toward genes that are disproportionately expressed in the central nervous system. Seeing that many of the genes in the education PGI are expressed in our epidermis (skin), musculoskeletal system, or fat cells would be a red flag signaling irrational evocation. But even if all the genes were in the central nervous system, they might affect things like our gross motor skills or our balance that shouldn't, in theory, determine educational attainment in modern society. Moreover, in most cases, there is no systematic way to automate the search for rational or irrational responses. It's just a smell test, case by case. This slipperiness is why population biologists and statistical geneticists get frustrated with us sociogenomicists: there are not universal biological mechanisms for characteristics that work through active or evocative gene-environment pathways (like selecting into college). That doesn't make the genes any less real in their effects though—it's just that those effects vary based on the social context.

The larger point is that almost all genetic effects that work through any behavioral or social mechanisms—most PGIs included—hold in a particular population at a particular time in history. A culture that sees red hair as desirable and attractive will obviously respond very differently to the genes for red hair (as expressed in the hair color itself) than a culture that sees red hair as, for example, the sign of the devil. The genes for eye color encode the same proteins in every country in all epochs, but the moment we are talking about a trait that involves any interaction with the world—whether that is what we eat, how people treat us, or where we choose to live—the effects of those genes become contingent on time and place.[23] For instance, the PGI for education, which was calculated with data from capitalist societies, did not predict years of completed schooling very well among subjects educated in Soviet-era Estonia but predicted schooling just fine for those born into post-USSR, capitalist Estonia. That doesn't necessarily mean that genes didn't predict educational outcomes very much in the USSR (though it could mean that)—it could simply mean that to make a predictive PGI in the USSR, you needed to choose different weights for different genes than in the U.S. or UK PGI. It could also be that different genes predicted educational success under communism by evoking different responses. Perhaps teachers prized sociality over individual expression, or rote memorization skills over critical thinking abilities.

This is especially true of traits linked with inequality. For instance, when calories were scarce and it was prestigious to be plump, the PGI for income might have picked up *positive* effects of being overweight—wealthy, heavier people might have been looked on favorably by teachers, employers, investors, and the like. The evocative effects of skin tone, of course, are tied up with the unequal relationships associated with colonialism, slavery, and racial stratification. And baldness was considered a sign of gravitas in some societies, such as ancient Rome, where age and wisdom were valued more than they are in modern U.S. society.

So, we work with the PGI we are given for the society we analyze, knowing that some portion of its effects probably work through undesirable or immoral genetic evocation. But aiming our PGI at different populations and epochs can help us narrow down what are the rational effects of our genes and what are not—as in the case of skin tone or male pattern baldness.

The bottom line is that, yet again, we cannot easily separate out nature from nurture. This fact is extremely frustrating from a scientific standpoint. Here we are with this amazing tool with which we can take a saliva sample from a baby and assay its adult height within two inches. We can predict that baby's odds of completing college. There is no limit to the number of traits we may be able to genetically predict using the DNA extracted from that saliva—but then, if something changes in our society, all bets are off. Moreover, even for *this* society at the *present* time, we cannot say exactly how much of the PGI's effect works entirely within our bodies through the actions of, say, enzymes; how much is dependent on how we engage with the environment (active GE correlation); and what percentage results from the social environment's response to our genes—both rational and irrational GE. It's like having an incredibly powerful telescope that can actually see dark matter for the first time and tell us its mass and effects on the universe but which still can't tell us what the hell it's made of.

JOYCE CAROL OATES ONCE SAID, "I DON'T CHANGE, I JUST BECOME MORE myself." Whether or not we are the same person over the course of our lives is an interesting philosophical question. In his reporting, *New Yorker* writer Joshua Rothman found that people are split on the issue. Some, like Oates, think they have always been the same person, while others cannot reconcile the person they were at eighteen with the mature adult they are at fifty. Understandably, those who have experienced discrete ruptures—the trauma of a war or immigration to a new land, say—are much more likely to say they have been different people over the course of their lives.

Genetics, and more specifically active and evocative GE, speaks to this philosophical conundrum. We might think of ourselves as grand buildings, designed by an architect who laid out the plans in a four-letter code called DNA. As construction proceeds, we look more and more like those plans intended us to. Indeed, while being built, we may bear little resemblance to the final product, with exposed I-beams, rebar poking out of cement columns, and scaffolding all around us. But gradually, we start to look like that 3-D rendering imagined from the 3.2 million nucleotide base-pairs that encode us. Sure, adjustments are made along the way. Perhaps a contractor tried to save money and skimped on the concrete. Or a certain cantilevered balcony won't work as designed because the architect underestimated the wind shearing at the upper levels. Or perhaps substitutions for flooring material need to be made because the proposed wood cannot be sourced.

These alterations of the architect's plan are analogous to the environmental impacts that affect how we turn out. For some of our outcomes, building construction is an apt metaphor for the relationship between our DNA and who we become—whereby active or evocative GE plays little role. Adult height works this way—there may be environmental shocks along the way, like illnesses or nutritional deficits, that make us not reach our full genetic potential, but otherwise we are built according to spec regardless of the environments we seek out or evoke in response to our genes. Indeed, most traits fall into this category: the environment matters, but it merely adds or subtracts from what our DNA would otherwise do in its "optimal soil," so to speak. Construction begins and ends at different times in our lives. Adult height is a building that is erected in the first twenty years of life. Male pattern baldness is something that begins in a man's twenties, perhaps, and continues for the rest of his life. A woman's fertility begins around age twelve and ends around age fifty. Notwithstanding some small amount of shrinkage that happens with osteoporosis, if I measure the height of someone when they are twenty-five or when they are forty-five, I am going to get the same answer. More importantly, the genetic contribution to height at twenty-five and height at forty-five is the same—after all, construction is done. What I mean by this is that variation in genes in a population will explain the same proportion of variation in adult height regardless of the age when the measurements are made. Most of the examples that fall into this bucket are physiological traits. But that is by no means a hard-and-fast rule. The key, from a sociogenomics point of view, is that once the trait itself is fully realized, the genetic contribution appears the same no matter when we measure it.

But for some human characteristics, the building analogy falls apart. Rather than a set plan for a building of a given height, depth, and width, the DNA instructs the organism to react to the environment. Intelligence and political attitudes fall into this category. Instead of laying out a plan for a brain to develop into a genius or a dolt, a liberal firebrand or a conservative, DNA says: here's a machine learning approach that I've embedded into your hardware; now go out and get some training data. (Never mind whether machine learning actually replicates how the human mind develops, it's more of a metaphor here.) What those training data are affects how quickly the AI (or rather NI—natural intelligence) learns, what it learns, and even whether it *wants* to collect more data to further refine its algorithm. Sure, some machine learning approaches (and the hardware they are run on) outperform others and learn faster given the same opportunities, but those differences slowly emerge over time and exposure to more and more opportunities to collect data on which to fine-tune the model. Some of those data fall into the lap of the NI genetic algorithm—those are the random environmental influences we all experience. But other information that the curious machine swallows up comes to it because the NI sought it out (active gene-environment correlation) or performed little experiments by, say, asking questions or testing the boundaries of parents (evocative gene-environment relations). The end result is that, like Joyce Carol Oates, the more complete the exposure to the environment, the more our hardware matters, and the more its essential potential shines through. Without comprehensive training data, the worst and the best AI or NI models will be suboptimal because of the limitations of the information they happened to have encountered.

This distinction between genetic building specs and genetic AI is not simply a matter of one characteristic—say, adult height—being less affected by the environment than another trait—say, intelligence. After all, the heritability of adult height is only slightly higher than that of adult cognitive ability (80 percent vs. 75 percent). And it's not that our test scores are constantly changing over our life span (though they do decline at older ages). It's that the process by which the genes express themselves is different with respect to the environment; how they have their effect is dependent on how we interact with and extract information from the environment—that is, active and evocative GE. In other words, it's how that 80 or 75 percent comes to fruition. For some outcomes like height, the environments that being initially tall or short lead us to (like the basketball court, perhaps) do not actually affect the ultimate outcome—adult stature. Our presence or absence on a basketball court is not an important part of the causal pathway for height-affecting genes to have their effects on height. But for something like cognition, the active and evocative environments genes engender are critical, necessary steps in the path from DNA to test scores.

But, ironically, for these psychosocial traits, the more the natural intelligence apparatus interacts with the world, the less important are the *random* aspects of that world that happen to us. What I mean by this is that when we measure something like intelligence, we find that early in life the environmental component is large—on the order of 55 percent at age five. But by age thirty-five, that proportion has shrunk to 20 percent. Conversely, the genetic component becomes more and more predictive as we age into adulthood. At age five the genetic component is 45 percent but by thirty-five it has ballooned to 80 percent. Many, but not all, social and behavioral traits work this way.[24] It's kind of like a fuzzy picture coming into focus. We don't know the exact extent to which active gene-environment dynamics are responsible for this process unfolding as compared to evocative gene-environment dynamics.[25] So, I guess if we think of our inner core as represented by our DNA, Oates is right: the more we interact with life experiences, the more we come into focus. But that's only made possible by "training" with our environmental interlocutors.

7

The Genetic Prism

Festooned with mustard-yellow drapes and a dangling American flag, the room resembled a Grange hall on bingo night. At center stage sat a wide vase containing 366 blue plastic lotto balls shaped like capsules, and over that vessel stood Representative Alexander Pirnie of New York. As his hand dug into the vase for a blue capsule, he averted his eyes, like a game-show contestant pulling prizes from a mystery bag. While the number of U.S. television viewers didn't nearly reach the level of the Apollo 11 moon landing a few months earlier, the consequences for U.S. families were much greater. It was December 1, 1969: the first drawing of the Vietnam Selective Service Lotteries. Until then, draft status had been decided by local draft boards; President Nixon had changed all that in an effort to make the system fairer.

Inside each capsule was a small sheet, to be pulled out like the slip from a fortune cookie. On each slip was a birth date corresponding to men born in the years 1944–1950. (Nineteen-forty-four and 1948 were leap years, hence the 366th capsule.) The first date Pirnie pulled out of the vat was September 14. Men born on this day would be the first called up to military service. After this ceremonial draw, the duty was taken over by students. (Nixon had wanted to show off young people willing to engage with the draft.) The last number called that night corresponded to men born on June 8. Over the next year, the first 195 dates selected ended up being called up, and those with birthdates assigned numbers over 195 were exempted from compulsory military service.

Jim, a college student who received a draft number of two in that 1969 lottery described the evening: "I remember the bar where we started drinking, but the march down State Street will be forever lost in a fog. Being No. 2, whenever the cry went up for lottery numbers, I was always the winner, and the beer was free for me." But free beer that night was hardly recompense, as described by another student, Gary: "everyone in the building crammed into my room to watch the lottery on my old black-and-white TV. Someone brought a six pack of beer that was to be awarded the 'winner,' i.e., the person having the lowest number of all the guys in the building. Well, unfortunately I won the six pack with No. 23. It was such an insignificant prize for something so potentially awful."[1]

The consequences could not have been more real. "I found out my number was 303," Peter described:

I had lucked out. One of my closest friends, Glen, wasn't as lucky. His number was 36. But Glen was the eternal optimist. I'll always remember his reaction: 'So I'll go . . . and I'll come back.' He was the first on the floor to go to Vietnam. He wrote me often from his outpost . . . and even sent me back one of his green army shirts with his name sewn on above the breast pocket. I went on to be a correspondent for *Newsweek* and covered the war at home. In the end, Glen kept his word. He went to Vietnam. He came back, and thankfully in one piece . . . physically. But mentally he was never the same . . . he simply stopped writing and disappeared. I still have that shirt today, a reminder of how the lottery changed both our lives.[2]

As unfair as it might have seemed to arbitrarily put some young men's lives at risk and spare others based on the randomness of a pile of plastic capsules, the lottery was thought to be a vast improvement on the prior selective service system, in which men enrolled in college earned deferments and local draft boards enjoyed enormous discretion to exempt individuals from conscription. Indeed, this rationalization of the process through random selection was an astute political move by the new Nixon Administration. While it did little to blunt overall dissatisfaction with the Vietnam War, it was popular in eliminating the apparent corruption inherent in the former system. (It is another question altogether why a congressman would think it would be good publicity to be the one to pull out the first number associated with being sent to fight in the jungles of Southeast Asia.)

As the Vietnam War limped along, there was a new draft lottery each year—though they were not televised like the first one. The draft classes of 1970, 1971, and 1972 were called for service, but as the U.S. role in the war wound down, future drafts were moot. The draft lottery was suspended after the 1975 drawing.

The Vietnam War occupied a special place in my childhood. I lived through it as a very small child—too young to know it was even happening but becoming very aware of its aftermath. Just as countless baby boomer kids played World War II in the 1950s, during the 1970s, I pretended the stand of bamboo near my grandparents' house was actually growing in the Mekong Delta. But besides my choice of games, I could feel the war's impact all around me. Many of the homeless people I passed each day had cardboard signs proclaiming that they were Vietnam vets. Many of the heroin users I saw on the street or on the subway were returnees self-medicating their trauma—many of whom had been sent there randomly by the lottery.

It wasn't just me, of course. The whole country was reckoning with the aftermath of the war—from the rise in drug addition, to Agent Orange–induced cancers, to the rash of suicides and other deaths of despair among veterans.

Even though the hardships faced by Vietnam veterans were plain to see, not everyone who returned from Southeast Asia experienced them, of course. Many vets managed to kick whatever bad habits they may have developed in the theater of war—smoking, pot, alcohol. Most settled down and started families. Others took advantage of the G.I. Bill to get more schooling.

The 64,000-dollar question, then, is why were some young men's lives devastated by the war while others escaped relatively unscathed? The individual differences between draftees' combat experiences might explain some of the differences in who, for example, ended up nursing a tobacco or alcohol addiction for the rest of their lives versus who managed to leave substance usage behind in the jungles of Vietnam. But that's probably not the whole story. We have long known that people react to environments differently, even when those environments are exactly the same. Plenty of people react quite differently to much less dramatic environments: Some people quit smoking when the surgeon general issued his famous 1964 warning on its health effects; others kept puffing away. Some folks take advantage of new schooling opportunities, and others don't. And some bodies balloon when they are faced with the veritable smorgasbord of unlimited calories that modern U.S. society offers while others manage to remain thin. One man develops asthma in a city with poor air quality while his brother shows no signs of a reactive airway. Some veterans from a platoon develop PTSD after combat while others from the same unit seem to escape war unscathed. And Leila, from my neighborhood, develops agoraphobia while Crystal ends up a scientist. There are a hundred different possible responses to a single environment, such as a math class, a war, or a childhood of poverty. But for a long time, the different ways that individuals responded to environments—poverty or fortune, war or peace, isolation or social immersion, to name a few—appeared random.

AS WE'VE SEEN, GENES GUIDE DIFFERENT PEOPLE TO DIFFERENT environments—but they also determine how we *experience* that environment once we're in it. If the roots of how people are affected differentially by an environment have to do with their genes, we call that a *gene-environment (GE) interaction*. A gene-environment interaction means not just that both your genes and your environment matter for a given outcome but something more: how the environment affects you depends on what DNA you hold in your genetic deck of cards. Similarly, the effects of your genes depend on what environments you experience: if you have genes for aggression, and you're raised in a violent neighborhood, you may end up in prison, but if you are born to great wealth with those same genes, you could wind up as a successful CEO. That would be one major GE interaction.

Since the discovery in 1953 of the GE interaction with respect to PKU and diet, there have been thousands of studies on others. Some are in the medical domain: smokers with a certain version of the NAT2 gene are particularly prone to bladder cancer. The genes for skin tone interact with how much sun someone gets to affect both their risk for skin cancer, on the one hand, and vitamin D deficiency on the other. State-level tobacco taxes interact with smoking genotypes so that some people are highly discouraged from puffing on cigarettes by high taxes while others with a more addictive genotype make choices that are unaffected by those same taxes. Compulsory schooling laws that force students to stay in school for longer attenuate the effects of genes on education, even beyond the law's minimal level, leading more people to obtain college and graduate degrees. In my own work, I have shown an interaction between identical twins' depression genotype and how much nutrition they receive in utero as measured by their birth weight.[3] Which twin got more nutrition is random thanks to where they each attach to the placenta, so it's purely environmental. The list goes on in both medical and social science. Now that we finally have the ability to measure genes, it seems the rule, rather than the exception, that they impact people differently based on their environmental exposures.

Pulling back from these specific examples, we can observe that, *in general*, some genotypes tend to generate large responses to environmental changes, while others keep the ship steady no matter what kind of storms hit it. In a commonly used metaphor in the field, some people are orchids—if given the exact right conditions in terms of humidity and temperature, they can bloom into some of the most beautiful flowers on Earth, where each one is uniquely stunning. But with suboptimal conditions, they wither, the petals fall off, and they quickly die. Other people are dandelions. Dandelions all appear to be more or less the same. They are fine to look at, but nothing to write home about in the world of flowering plants. Yet they always manage to attain that minimal level of blooming. Dandelions grow in sandy soil. They can survive in the arctic. They blossom in almost any biome on the planet where plants can grow. In the same way, some people have genotypes that predispose them to respond dramatically to certain environments, while others have genotypes that cause them to be relatively unaffected by those same environments. Fortunate human "orchids" end up more beautiful (i.e., healthier and more successful) than the plain old dandelions, which have a "ceiling" of achievement; but the likelihood of an orchid withering on the vine (epic fail) is greater. It's a high-risk, high-reward genetic strategy.

The fact that the effects of genes depend on environments and the effects of environments depend on genes takes us past the social genome and past gene-environment correlations to a total merger of nature and nurture. Literally, we cannot speak of a genetic effect absent the environment nor an environmental effect net of genetics.[4] They form that single Möbius strip.

Thus, if we want to grasp why people end up where they do in life, then it's critically important that we understand how gene-environment interactions operate across health, development, education, marriage, and so on. It's not enough to know how much genes contribute to average outcomes and ditto for the environment. To get a complete picture of how society works, we need to be able to describe how genes and environments interlace with each other, affecting each other's effects, if you will. If we can't know for certain that Leila and Crystal reacted differently to their childhood neighborhood because of their genes, our ability to understand people's outcomes in modern society (much less predict them) remains limited. Who "wins" is not just a matter of which genes someone carries, or which environments they experience; it's the unique combination of those two factors. But, like my father wagering an exacta rather than a straight win bet, it's much harder to get GE interactions right—they involve properly assessing the genetic inputs as well as the environmental ones.

SCIENTISTS HAVE LONG KNOWN THAT PLANTS AND ANIMALS RESPOND differently to environments based on their genes. In the lab, they can experimentally manipulate the genomes of these "model" organisms as well as the environments they experience.

When we turn to those pesky, free-range humans, however, it becomes extremely difficult to prove the existence of a GE interaction—despite their apparent ubiquity. That's because it's very tricky to distinguish between non-causal gene-environment *correlations* and causal gene-environment *interactions*. Some genes and environments tend to cluster together, but their particular combination isn't what is having important downstream consequences. For GE interactions, the unique combination of a given environment with a particular genotype results in sequelae we can observe. For some medical traits—like the interaction between smoking and the NAT2 gene with respect to risk for bladder cancer—we can confirm GE interaction effects. We can experiment on animals in the lab, where we can eliminate confounding GE correlations by design because both the genotype and the environment are totally controlled. We can also hope to break down and understand the exact molecular process happening. But for other traits—like the relationship between cigarette taxes, genotype, and smoking—it becomes more difficult. We can't, after all, subject mice to different taxation rates and see whether their smoking habits are altered. Among humans, for most outcomes, detecting true GE interaction effects was out of reach until recently. It took nothing less than a revolution in the social sciences to match the molecular genetics revolution in order for us to be able to accurately distinguish GE correlation from GE interaction.

That didn't stop researchers from asserting that there were GE interactions left and right—merely by taking a gene and showing that its effects appeared to be different in two scenarios. In fact, the first study ever published using molecular genetic markers to examine how genes and environments combine to produce mental health and behavioral outcomes in humans suffered from this conflation of GE correlation and GE interaction. In the paper, published in the prestigious journal *Science* in 2002, Avshalom Caspi and colleagues showed that men with one version of the MAOA gene were more likely to grow up to commit antisocial behaviors *only* if they experienced childhood abuse or neglect.[5] The MAOA gene was the gene, and the childhood abuse/neglect was the environment—their results seemed to point to an important *gene-by-environment interaction effect* (*GxE*).

But what if another gene was at work, one the researchers didn't measure? Even in the age of GWAS, we are still only assessing commonly varying genetic loci, so it's always possible that some new mutation or other rare gene could be the most important actor. Let's call this unmeasured gene "Z." Imagine that parents who are abusive or neglectful tend to have both the MAOA gene and this mysterious, unmeasured gene Z themselves because it takes both to be a truly bad parent. Now imagine that the parents have passed both these genes on to their child. The parents, in other words, are providing both an environment (abuse or neglect) and *two separate genes* (the MAOA gene and gene Z) to their child.

It's possible that the childhood environment did indeed interact with the MAOA gene, tipping the child over the edge into antisociality. But it's also possible that the environment had nothing to do with it and that, rather, it was the fact that the child inherited both the MAOA gene and our mystery gene Z that led them to commit antisocial acts. Since we didn't measure the presence of gene Z explicitly, parental abuse/neglect (if it is due to the presence of the MAOA gene and gene Z in the parents) is the best approximation of whether both genes are present in the offspring. So, in this case, it could be the interaction of two genes (*gene-by-gene interaction* [GxG], called *epistasis* in biology) that is causing the antisocial behavior in the child—and the so-called gene-environment interaction might be only a mirage.[6] Or, alternatively, it could still be a gene-environment interaction, but between gene Z and parental abuse; we would misattribute it to the MAOA gene because we failed to measure all the risk alleles. A final possibility is that the combination of the risky genes in the parents cause their behavior to be abusive, this in turn causes offspring to become antisocial, while inheriting MAOA (or Z) alone has nothing to do with it.

This is a classic case of passive gene-environment correlation: the mystery gene Z is correlated with the childhood environment of abuse or neglect. The genes of the child are correlated with its environment because the genes, when inside the parents, shaped that parental environment. That's not to say that it's not possible that there is a GE interaction between parental neglect and a particular gene like MAOA in the child; it's just that we cannot separate out gene-environment interaction from gene-gene interaction in the face of passive GE (or any other type of GE correlation). Since we cannot measure all possible genes, there could always be a gene Z lurking behind the environment in a gene-environment interaction. Unless we can find some way to rule out that possibility, we can't be sure that what we're seeing is a true gene-environment interaction.

Take college, for instance. We know from our research that our genes lead us toward (or away from) college through GE correlation. We also know that among people with the same educational PGI, those individuals who actually attend college have higher incomes on average, so there must be a real, environmental college effect.[7] But to identify a gene-environment interaction, we're asking not whether there is an additive effect of going to college (say a 10 percent wage boost, on average, for all PGI levels) but whether there is a *multiplicative* effect (such as someone with a high PGI getting a 50 percent wage boost and someone with a low PGI getting no boost). Do students with high education PGIs that sailed into their dream school "get more" out of college? Or is it the students with lower PGIs who managed to defy their genetic odds to get admitted to the same school who actually have the most to gain? That's the kind of question at the heart of gene-environment interactions.

It might seem that we can simply use the education PGI to group people by their relevant genotypes and see if that explains who gets a big bump from a college degree and who doesn't. But even if we do find that people with a certain PGI for education get a bigger bump from college, this isn't enough to prove a gene-environment interaction. The problem is that the PGI for education (or for any other trait) doesn't capture all our genes—it captures only the commonly varying genes that are measured in a GWAS as they relate to education.

There could be rare genes—or even common, health-related genes or entrepreneurial genes—that college attendees tend to have that multiply the effect of the education PGI on income.

This puts us in the same bind in which Caspi and colleagues found themselves with respect to parental abuse or neglect and MAOA: we can't rule out the existence of some pesky gene Z that isn't captured by the education PGI but is, nevertheless, pushing us toward college enrollment through gene-environment correlation. So, if people who have a high PGI for education get more economic returns from a college education than those with a lower PGI, it could be that it's a gene-environment interaction, but it could also be that other genes, not captured by the PGI, that also drive someone to college are what interact with the education PGI to amplify its downstream effect on wages and college attendance. In this case, the environment of college would be acting as a stand-in or substitute measure for some unmeasured genes that are amplifying the effect of the education PGI, so it's not really the effect of college at all, but the unmeasured genes that got the students to college. The result is that, as in the case of Caspi's and colleagues' papers, we cannot know whether we are seeing GxE or GxG at work when different levels of a PGI seem to benefit more or less from a college degree.

The point is that the PGI (or any genetic measure) alone is a necessary but not sufficient tool for studying gene-environment interactions. Using only PGIs, we can't prove that a GE interaction explains why some veterans get PTSD and some don't, or why certain people get more out of college than others, since there could be some other phantom PGI that predicts "joining the military" or "getting the most of college," and it is the multiplicative effect of those PGIs on the PTSD or education PGI, respectively, that we are errantly detecting. Once we recognize that among us free-range humans, genes drive so many of the environments we experience, this problem becomes manifestly visible and rife. We need some other tool, and since we

cannot measure all possible genes absent the whole-genome sequencing of massive numbers of people, nor can we create all possible PGIs,
we can only get out of this trap if we have an environment that we
know for a fact is unrelated to genes.

FOR DECADES, A SIMILAR DILEMMA PLAGUED ALMOST ALL OF SOCIAL
science. Since environments are not randomly assigned to free-range
humans, how can we know that the putative environment we are pur-
porting to measure is the actual cause of the outcome we are studying, as opposed to the genes that predisposed their bearers to seek
that environment? How can we know—going back to the example of
schooling—that an additional year of school or a college degree has
such salutatory effects when very different people graduate and don't
graduate? It could be the nature (or prior nurture) of those kids that
makes them go to college *and* that raises wages. Is it selection (who
goes) or treatment (what happens on campus)? Likewise, if veterans
suffer from worse health or earn less money after serving, is that due
to the treatment of war or the selection into the military of people
with different genetic dispositions? This problem of distinguishing
selection from treatment is not just a problem for gene-environment
interaction research; it's, as I mentioned, the fundamental problem of
all observational social science—but it becomes particularly acute in
the case of GxE studies.

The solution to this ever-present conundrum is to find an environment that we know isn't related to our genes—the ones we measure
or the ones we don't measure. What we really need is an experiment
that randomly assigns environment A to Group 1 and environment B
to Group 2. Otherwise, there's no way to separate treatment effects
from selection effects.

But with a few exceptions—like social psychology studies run
in the basement of the psych building—social scientists can't run

experiments on humans. At least not meaningful ones. We cannot randomly assign some couples to stay married and others to divorce. We can't make some people religious and others atheistic. And, of course, we can't make some people Black and others white. Only with a controlled experiment can you know for sure that one factor is causing another. If I give some families an extra thousand dollars a month and give nothing to other families from the same population, then I know the money is causing whatever outcome I am measuring since, as long as I flipped a coin to determine which families got the cash, and the population is large enough that all other relevant random factors average out, they are, on net, no different from each other, so nothing else could be causing any observed differences between the two groups.

But what can we do when we can't flip a coin to assign people to their fates? For many outcomes human scientists care about, it's either infeasible or unethical to randomly assign people to treatment and control groups. As mentioned in Chapter 1, it's not ethical to assign some people to drink more liquor to study the effects of ethanol on the body. In the case of family income, there have been, in fact, some studies that randomized households to receive extra money (or not); but as you can imagine, this is a very expensive study to run for any length of time at a large scale.

A seminal 1990 paper by labor economist Joshua Angrist pointed toward a solution when explicit experimentation is not possible. Angrist won the Nobel Memorial Prize in Economic Sciences in 2021 for his development of what became known as the natural experiments school of economics (along with David Card and Guido Imbens). While I had talked about hopping freight trains after graduating college, Angrist was already doing it in high school, albeit in Pittsburgh rather than the American West. While I enrolled in ROTC in college but quit before signing on the dotted line to commit to a commission as a second lieutenant in the U.S. Army, Angrist initially forsook an offer to attend Princeton for graduate school in economics in order to become a paratrooper. After his military service, Angrist finally accepted the offer of admission to Princeton's PhD program that had been extended by Professor Orley Ashenfelter on the strength of his undergraduate thesis (which he had read as a visiting evaluator at Oberlin College, Angrist's alma mater).

One day, during a graduate school class, Ashenfelter mentioned a study from 1984 that had used the Vietnam draft lottery to explore the post-war mortality of the soldiers who had served in the military.[8] Since the draft numbers were randomly assigned, they offered a way to figure out the true effect of military service on mortality without bias from other factors that might lead someone to enlist and which, in turn, might independently affect post-war mortality. "Orley said, 'That's such a great idea—somebody should do that for [former soldiers'] earnings,'" Angrist recalled to *MIT Technology Review* magazine. He "got to work that afternoon."[9]

Angrist had realized that the draft lottery was a *natural experiment*—an unintended event that randomly exposes some people to an environment and not others. Each fortune-cookie-sized strip of paper that Pirnie and the students pulled out of the capsules on December 1, 1969, scheduled the assignment of what scientists would call a *treatment condition*—an intervention that, from that day onward, would alter the life outcomes its subjects experienced, just as a pill randomly allocated in a pharmaceutical trial might alter a participant's health. This procedure made those with lower numbers more likely to face military service, not because of any personal attribute likely to be correlated with life outcomes, but because of a random draw of an innocuous attribute unrelated to much of anything—their birthdate. That was its intention: with induction assigned by chance, no correlation *should* exist between service and inductees' personal attributes (social class, race, risk tolerance, and so forth). This was critical for Angrist; even though he was not concerned with genes, he still had to be certain there was no correlation between military service and any other personal attributes of veterans and non-veterans. With the correlation broken by the draft lottery, he could see the "pure" environmental effect of being drafted. In other words, randomness ensured that any other salient variables would have the same distribution in both groups and therefore would not affect average outcomes.

Angrist's dissertation, which became that seminal 1990 paper, did just what Ashenfelter had suggested.[10] After wrangling a lot of data from old government mainframes, he used the Vietnam Era draft lotteries to assess the impact of military service on later earnings. Angrist found that serving in the military during the era of the Vietnam War resulted in a 15 percent wage drop for whites over the next decade. Black inductees, by contrast, suffered no such disadvantage despite having a greater than average likelihood of exposure to combat and other less desirous aspects of life in the army. The putative theory behind this race difference was that during the 1970s Black Americans faced such discrimination and limited opportunities in the labor market that losing two years of civilian labor experience did little to handicap them any further than they already were. Put another way, if you belong to a group that's largely confined to flipping burgers or delivering packages thanks to racism in the job market, then you will be flipping burgers all the same after you're drafted as before, or if you were never drafted at all.[11]

The 15 (or zero) percent treatment effect of military service itself was not so revolutionary—it wasn't entirely unexpected or novel. It was how Angrist got to the result. Because the Vietnam draft lottery was an accidental experiment, Angrist showed, as definitively as possible, the causal effect of military service on wages.[12]

Angrist's paper launched what is now called the credibility revolution in economics and the causal revolution in social science more generally—of which sociogenomics can be said to be an heir.[13] Economists had long been aware that if you observe that factors A and B move in tandem, that doesn't necessarily mean that A caused B, as you may assume in your theory. It could be that B caused A or that a third factor, C, caused both. When you aren't running an experiment that changes A to observe what happens in B, it's pretty hard to figure out what's causing what. Correlation is not causation, as the old social science adage goes. Take inflation and wage growth by way of example. Whenever inflation is high, wages tend to be rising fast, too. Maybe rising wages are the main cause of inflation. Or, maybe inflation is causing employees to demand higher wages. Or it could be that a third factor, like a global oil shock or the Federal Reserve's monetary policy, is causing both wages and inflation to rise. In the case of inflation and wages, all three possibilities are plausible; but figuring out the exact contribution of each to the wage-price spiral, as it is called, is tricky. Angrist showed his colleagues that there was a solution to this conundrum: find an appropriate natural experiment to disentangle what's going on.

After Angrist's paper, finding natural experiments beyond the draft lottery became the quest of thousands of academic economists around the world. As it turns out, there is plenty of randomness in life when you are looking for it; natural experiments abound in the social world. When a judge suddenly orders an end to school bussing—as happened in the Charlotte-Mecklenburg district in North Carolina in 2001—that generates a natural experiment on the effects of bussing, school district zones, and racial integration (bussing reduces racial inequality in multiple outcomes).[14] Or, sometimes the experiments are literally accidental, as when the 1986 Chernobyl nuclear power plant disaster in the Soviet Union caused Swedes in various parts of the country to experience different levels of radiation through iodine-131 exposure from the toxic cloud that floated north-west from Ukraine (those exposed in utero suffered from cognitive deficits and were less likely to graduate high school).[15] Other times natural experiments are even, well, natural. The weather—when it is unexpected—provides a random treatment. In this vein, some scholars used random variation in the weather to study the effect of drought—and associated crop losses—on the likelihood of civil war in a country (no surprise here: crop losses increase the likelihood of armed conflict, as they did in Rwanda in 1994).[16]

Scholars following in Angrist's footsteps have used the random assignment of cases to more or less lenient judges to estimate the effects of incarceration on later criminal behavior (jail turns out to be worse for recidivism); the timing of Ramadan to examine the effect of fasting on birth outcomes (lower birth weights in offspring); the effect of additional schooling imposed by changes in minimal school-ing laws (longer life expectancy); the presence of a boy with a femi-nine name in a classroom to study the effect of disruptive peers (à la Johnny Cash's hit, "A Boy Named Sue," worse test scores due to classroom conflict); the number of rivers that happen to slice through a state to examine the impact of school district size (more rivers chop-ping up a state equals smaller districts, which results in greater com-petition and, in turn, better schools); and hurricane devastation of the coffee crop in Brazil to study the effect of coffee prices on parental labor supply in Colombia (parents earn more money, but their kids are worse off due to less attention)—just to name a few.[17]

In 1996, as the credibility revolution swept through the field of economics, I had finished my PhD in sociology and found myself at my old stomping grounds, UC Berkeley. Gradually, I became aware of all the ways sociology could learn from the work going on next door in economics. Both fields were inhibited by the inability to run controlled experiments on free-range humans; but by hunting for and gleaning data from natural experiments in the domains we sociologists cared about—fertility, marriage, peers, social networks, and so on—we might generate more accurate and durable findings.

Toward the end of my time at Berkeley, I was presenting some of my postdoctoral research on birth weight to my fellow fellows, some of whom were economists. Poverty, lack of education, and ill health have always been linked. In the U.S., about one in twelve babies are born under five pounds, eight ounces, the threshold for being classified low birth weight. An underweight baby can result from either having been born too early or growing too little in utero. Both these causes can be influenced by the uterine environment, which, in turn, is affected by the larger environment a pregnant woman is experiencing—including the stresses of poverty. Being a small baby, in turn, is associated with a whole bunch of risks. Some are immediate—like higher rates of infant mortality. And some are longer term—like developmental delays, shorter adult height, poorer socioeconomic prospects, and even lower life expectancy.

But if poor moms tended to give birth to smaller babies, on average, how could I disentangle the effect of growing up poor from having been a low-birth-weight baby? In other words, was maternal poverty the driver of both low birth weight and what happened to those babies, or did a baby's weight at birth actually help cause the intergenerational cycle of poverty and ill-health? If it was all the mom's poverty that mattered—particularly during pregnancy—society could gain a lot by helping to buoy the incomes of poor (pregnant) women. Not only would it improve the health of many women, maternal income support might actually save money on later healthcare costs as those kids grew up. Conversely, if birth weight was a key cause of adult problems—ill-health to poor education to economic dependency—then if we intervened to improve birth outcomes, this would be the most efficient way to improve health and economic self-sufficiency.

I had already become aware of the problems of unobserved variable bias with respect to studying poverty thanks to Susan Mayer's book *What Money Can't Buy* and knew that the study of birth weight faced the same threats. *Unobserved variable bias* is the idea that I've mentioned a few times already: the notion that some unexpected factor C could be driving both A (birth weight) and B (adult outcomes like poverty). If we don't somehow factor out that third, unmeasured force, then we end up with wrong results about the relationship between A and B. If we didn't factor out, for instance, the health behaviors or genetics of the mom, we might wrongly conclude that birth weight had all sorts of effects when it was really a mom's drinking while pregnant that caused low birth weight and a bunch of other bad outcomes that manifest later in the child. If we don't factor out the genetics of the mom, it could be the case that her genes (and those she passes onto her child) are what is driving everything and all our policy efforts to improve birth outcomes would be for naught.

To address the issue that low-birth-weight babies more typically came from families that suffered from a whole host of other disadvantages—economic, health, racial—I contrasted siblings from the same family, one of whom was heavier than the other at birth. What I found was that, indeed, even comparing full siblings born to the same mom—and factoring out any differences of gender, birth order, maternal age during the pregnancy, family income fluctuations—a child who was born at a normal birth weight was 75 percent more likely to graduate from high school on time than a sibling who was born at a birth weight of less than five pounds eight ounces. I was excited about this result because it suggested that we could improve educational outcomes and reduce disparities (by income and race, in particular) if we could figure out how to raise the weights of these small babies.

When I presented these results, an economist named Michael Greenstone (who later served on Obama's Council of Economic Advisors and who is a leading environmental economist at the University of Chicago) asked if I'd considered using the roll-out of the

Women, Infants, and Children (WIC) nutrition program, launched in the early 1970s, as an instrumental variable for birth weight. I was too embarrassed to ask what an instrumental variable was.

An *instrumental variable (IV)*, it turned out, is a factor that causes a natural experiment by shaking up another factor. In that original 1990 Joshua Angrist paper that launched the revolution, the randomly drawn draft number for each nineteen-year-old man was the IV; that number, in turn, affected the cause Angrist wanted to study—military service—without messing with other outcomes of interest directly (in Angrist's case, wages).

The natural experiment approach could indeed be useful to my research on birth weight. My sibling comparisons eliminated a lot of other competing factors, but they left a critical one in place: the genetics of each baby. It could be that the genetic differences between the low-birth-weight sib and the normal-weight sib were those "unobserved variables" driving both their birth outcomes and the inequality in their adult outcomes. An IV, or a similar tool from the economics toolbox, might help me eliminate the potential influence of genetics that still plagued my research. Since an instrumental variable randomizes a treatment—say a year more of school or having to serve in the military—it's unrelated to the genetics (or anything else) of the people who get or don't get the treatment. IV skirts the whole issue with free-range humans—that they often select their environments (i.e., treatments)—thereby taking genes (and other pesky factors) out of the equation, *literally*.

As it turned out, for various reasons, WIC would not make a good IV for what I was trying to measure, but the larger lesson was not lost on me: I needed to find natural experiments that randomized whatever I was interested in studying.[18] I returned to Berkeley and pestered my office-suitemate, another economist, with constant questions as I tried to make sense of the more advanced math that economists used in their methods textbooks. My two-minute introduction to IVs in the hallway of a conference center opened my eyes to all the ways sociology could learn from the work going on in economics.

I became an overzealous convert to the credibility revolution and turned into something of an intellectual import dealer: I found natural experiments that had been gently used in economics (mostly to estimate effects on wages) and brought them into sociology to examine noneconomic outcomes. Over the years, I studied (or rather stole) any natural experiment I could find. I analyzed whether having a girl or a boy as your first-born child affected your political views (girls made parents more conservative, particularly with respect to issues surrounding the regulation of sexuality).[19] I looked at identical Black and white twins who diverged on their birth weights to show that Black people suffered higher infant mortality rates thanks to post-birth treatment rather than pre-birth factors. With Gordon McCord and Jeffrey Sachs, I even studied climactic influences on mosquito reproduction to help understand the role of malaria-induced child mortality on economic development.[20] And, of course, I borrowed Joshua Angrist's original idea—using the Vietnam draft lottery—to examine everything from marriage and divorce (with Jennifer Heerwig) to public sector employment (with Tim Johnson) to long-term mortality (also with Jennifer Heerwig).[21]

Eventually I started to feel torn between the fields of economics and sociology, feeling like I didn't fit into either one. I wanted to retain the flexibility of sociology to ask far-reaching questions on basically any topic while aspiring to the methodological "credibility" of economists. Though I loved the detective game of finding and analyzing natural experiments, the approach suffered from the limitation that, like the drunk in a parable, I had to look for my keys under the lamppost only because that's where the light was shining. In other words, the research questions I could answer were driven by the availability of a natural experiment, rather than the importance of the inquiry. I was not alone in my frustration with the limited scope of what social scientists could investigate using the natural-experiments approach championed by the credibility revolution. According to a 2021 survey of leading economists by the London School of Economics, thanks to the credibility revolution, a majority felt that "researchers often seek good answers instead of good questions."[22]

The reason I had gotten into the field in the first place was not just to handicap humans in the great socioeconomic race as my father had handicapped thoroughbred horses, but to understand how accidents of birth shape who we become and how that process is, in turn, patterned by structural forces like rising economic inequality, racism, and so on. While in my early work I showed that parental wealth seemed to play a major role in perpetuating inequality, and I even envisioned a bold new social policy based on wealth rather than income support, the credibility revolution had eroded my confidence in my own conclusions. In science, you shouldn't move backward. Which is to say, knowing what I now knew about the limitations of my earlier observational approaches, I couldn't just throw up my hands and abandon the natural-experiment approach without acknowledging that other factors might be sneaking in and causing any differences I observed. Luckily, when I started investigating the genetic influences on children's outcomes in the late aughts, the natural-experiment framework that I had internalized the prior decade came in handy when I went to study the final piece of the puzzle of how we become who we are: gene-environment interaction effects. Natural experiments might constrain what you can study through their limited availability, but by merging them with genetic information, we come up with a whole new way to understand how nature and nurture work together to produce human outcomes. My emerging weariness of the credibility revolution was swept away by the possibilities offered by combining genetics and instrumental variables.

WHEN WE HAVE A PGI TO MEASURE GENETIC TENDENCIES *and* A NATU-
ral experiment like the Vietnam War, we have the two tools we need
to properly test for gene-environment interactions. This is what economist Lauren Schmitz and I did, revisiting the same Vietnam draft
lottery that Angrist had used to launch the credibility revolution.
We wanted to know if the PGI for smoking moderated the treatment
effect of being drafted during the Vietnam War on lifetime cigarette
use. We knew that the army was a very pro-smoking environment: it
was stressful; cigarettes were heavily subsidized and sometimes free;
and the social norms of masculinity encouraged smoking. But years
later, some people who were drafted ended up lifetime, heavy smokers and others didn't. It seemed random who was deeply affected by
military service when it came to smoking behavior.

Without the draft lottery, we couldn't have done our study. If we
had observed a big smoking rise among veterans with high smoking PGIs and no rise for veterans with low PGIs, it could still be the
case that unmeasured genes that cause people to enlist—say, markers related to risk taking—were activating the smoking PGI rather
than military service itself. The environment of the Army could just
be a proxy for risk-taking genes that enhance a genetic propensity to
smoke (i.e., the smoking PGI). But since we knew men's lottery status
affected their likelihood of serving, *and* we know that it was random
and unrelated to any genes, when we used draft numbers instead of
actual service, we knew we were measuring an environmental effect
that amplified genetic tendencies to smoke.[23]

When we analyzed lifetime smoking using the Vietnam draft
lottery as a stand-in for military service, we found that for people
with a low-smoking PGI, there was little effect of military service on
their lifetime smoking rates. They may or may not have smoked while
in the army, but if they did, they managed to quit afterward and not

keep the habit for life. But those with a high-smoking PGI ended up smoking a pack more a day for life compared to people with similar PGIs who hadn't been drafted. Keep in mind that we are not talking about the fact that those with a high-smoking PGI smoke more than those with a lower-smoking PGI on average whether or not they are veterans (which is also true); we are saying that those with a high PGI who happened to be drafted ended up in double jeopardy and smoked a pack more a day than those with the same PGI who were never called to service. What makes this a gene-environment interaction is the fact that your PGI determines how you respond to the environment. Those with a low-smoking PGI don't "react" at all to military service—at least in terms of post-war smoking. Those with a high PGI react a lot—they increase their smoking by a pack a day relative to their already-elevated rate of smoking overall. This, then, is a quintessential gene-environment interaction: a scenario where our genotype affects how we respond to a given environment. This paper represented one of the first "true" detections of a gene-environment interaction in human populations.

Getting drafted wasn't all bad news, however. Thanks to the G.I. Bill, those draftees who had a high PGI for education were more likely to graduate college than their counterparts with the same education PGI whose birth dates meant they were passed over by the draft board. But for those with a low PGI for education, it didn't make a difference whether they were drafted or not in terms of whether they graduated college. That represented another gene-environment interaction: those with a certain genotype reacted to the environment (in this case the college money they were eligible for) differently than those with an alternate genotype.

One way to think about the power of a PGI in combination with a natural experiment is to imagine a prism that refracts light. Everyone responds to an environment in a different way: in red, green, or violet wavelengths, say. When we just look at all the responses without the

PGI, the colors of the rainbow (the different responses) are mixed together, and all we can see is white light (i.e., the average response). For instance, when Angrist's original study found that there was a 15 percent wage penalty in the 1980s for getting drafted in the early 1970s (among whites), he was only seeing the white light. Some vets ended up very economically successful. Others floundered. When we lump them all together into one pile, we get the minus 15 percent average.

But when we use the prism of the PGI, we can view the entire spectrum of responses—and measure the brightness level of red versus green versus violet responses since they are now clearly sorted and visible by genotype.[24] If Angrist had had genotype information along with tax records, he might have been able to predict who lost 30 percent of their wages and who suffered no penalty for military service—within race. Indeed, our results on the education PGI and the effect of wartime service on college education suggests that Angrist would have found a genetic pattern to the income responses to military service. (We could not look at income in our study due to a comparatively small sample size.)

I've been talking about gene-environment interaction effects as if the PGI is the prism and the environment is the white light bent through that piece of glass to reveal different responses. However, when we see genes and the environment as the single-sided Möbius strip that they form together, it's just as accurate to say that environments modify the effects of genes as it is to describe GxE as genes moderating the effects of environments. In other words, in the case of the Vietnam smoking example, we can think of the PGI as the white light, and the draft number as the prism that bends the light. Namely, instead of trying to make sense of why some veterans smoke a lot and others don't by peering through the prism of their DNA, we could ask why some people with a high PGI for smoking don't smoke while some do. To see why a high PGI has different effects on people— after all, it does not predict smoking anywhere near perfectly—we can have a challenging environment (Vietnam) function as the prism to make sense of individual differences. The hell of war can refract innate genetic tendencies to smoke in a dramatic way. In the same vein, as hinted at before, a modern diet full of virtually unlimited rich foods can allow a thousand genetic flowers to bloom in terms of weight—refracting the white light of the BMI PGI into a whole range of body sizes. That said, since genotypes settled into the population over generations and the environments can change on a dime, in general I prefer to describe the GxE combo effect as genes moderating the impact of the environment.

The PGI prism often reveals gene-environment interaction effects most noticeably when a radical environmental change is experienced by an entire population—tobacco is brought from the New World; coffee cultivated in Yemen sweeps through Europe during the Renaissance; high-fructose corn syrup pervades processed foods after the passage of the 1980 Farm Bill. Sometimes, these widespread changes mean that traits that evolved in response to one environment become maladaptive in a new environment.

By way of example, some human genotypes that evolved in very hot climates with little sodium, like much of sub-Saharan Africa, are adapted to retain sodium so that the body has enough of this mineral to function properly. But take those genotypes, or rather the people with those genotypes, and plop them in a society where we get too much sodium (i.e., the contemporary United States with its highly processed diet) and throw in social stress due to racism to boot, and you end up with a group that suffers from elevated rates of hypertension (e.g., African Americans). A genotype that was highly adaptive in one context becomes maladaptive in another context.[25]

Likewise, genotypes that caused bodies to binge on whatever calories happened to be available might have been highly adaptive for most of human history when people didn't know when their next meal was coming. Better to eat as much as possible while the living was good and store any excess calories as fat to be burned later when times were lean. But today, in an environment of seemingly unlimited, uninterrupted, relatively cheap, and highly processed food (at least in some parts of the world), such genotypes lead to problems like obesity and type 2 diabetes.

Indeed, when colleagues and I examined the influence of the body mass index PGI over the course of the twentieth century, we found that the influence of genes on weight increased over that critical time period: those born at the beginning of the century lived most of their lives in a world of relative food scarcity; those born in the second half of the century (and especially in the last quarter) were presented with an environment where restraint was healthy. This is an example of a gene-environment interaction where the environment shifts so rapidly under the feet of the population, so to speak, that genotypes do not have enough time to adapt to the new world in which they find themselves.

Other times, genotypes may just vary by random chance under no evolutionary pressure but then face a novel environment, transforming a previously inconsequential variation in our genomes into a consequential one. Take smoking. The receptors for nicotine found in human brains predated humankind's exposure to tobacco with the discovery of the crop from the New World, first by indigenous Americans and then by Europeans. I can't say what those genes did before tobacco was widely available, but once there was an option to inhale smoke from tobacco leaves for a pleasant feeling, the genes that regulated that experience became very consequential as they predicted addiction. There is a gene-environment interaction in Europe pre- and post-1528, the year the Spanish introduced tobacco from the Americas. Before that year, the PGI for tobacco probably didn't predict adverse health effects; as tobacco use spread across the continent and the environment quickly shifted in Europe, however, that same PGI would have predicted lung cancer, emphysema, and shortened life expectancy.

But then the environmental landscape shifted again; this time due to information about the product, rather than the product itself. As described earlier, after the surgeon general's landmark report on the dangers of smoking was released in 1964, the PGI index for smoking predicted smoking better than it had before the report (and the overall heritability increased). This is a gene-environment interaction effect as well: if you came of age in the 1950s, your smoking PGI didn't predict your tobacco use very much. But if you came of age in the 1970s, say, the gap in smoking rates between those with a high PGI and a low PGI were substantial. That's because, as I mentioned, the information environment changed. Before the ills of tobacco use were widely known or accepted, many people kept the habit. Why should they quit if it was pleasurable and wasn't so bad for them? The U.S., in short, was a very pro-smoking environment, so many people smoked, regardless of genotype. But once new knowledge shifted the landscape, only those who were hard-wired to smoke were unable to give up the habit, and the PGI improved as a predictor. As the rational case for stopping smoking becomes stronger, genes play a more important role in determining which of us take up the habit (or fail to quit).

Similarly, the PGI for educational attainment is based on analyses of data of individuals from Western, capitalist societies. Thus, as mentioned earlier, it predicted poorly for individuals who attended school in Soviet Estonia. But when the USSR collapsed, and a new generation of students grew up in a more Western-style, capitalist schooling system, the PGI's predictive accuracy improved dramatically. The environment shifted and the way genetics mattered also changed—a gene-by-environment interaction effect.[26]

These radical environmental upheavals—from caloric abundance to the upheaval of a society's entire economic system—function like natural experiments just as the Vietnam draft lottery did. In these cases, however, the "control" group is the people who came of age before the big shakeup, and the treatment group consists of folks who lived under the new environmental regime. Since the environmental changes are not related to the genes of people who are in either group (evolution doesn't work that quickly), they satisfy our natural experiment criterion of being randomly assigned.

In general, we can begin to see a pattern here when we look at these large-scale "experiments": when new opportunities to make an informed choice arise thanks to a shift in the environment, genetics tends to predict better. There is a gene-environment interaction effect such that for those in the pre-change control group, people with the same PGI don't exhibit major differences. But post-change, the gaps between people with different PGIs swell—that is, genetic inequality rises. For instance, when we have the choice to eat unlimited amounts, the BMI PGI predicts big differences in weight; when we are all calorie restricted thanks to the environment, it doesn't. Hand-in-hand with rising overall BMI levels and rising inequality in BMI comes higher genetic influence/inequality. Conversely, when people were given better information to make a choice about whether to smoke or not, not only did smoking rates decline, but the genetic inequality in smoking rose. When women were given access to higher education, many took advantage of that opportunity and overall college graduation rates rose among U.S. females. But who tended to take advantage of that new educational option? All else equal, it tended to be women who had higher education PGIs—hence its increased predictive power. Opportunity and choice increase the inequality in outcomes due to genetic differences.

By contrast, when the environment is equalized through fiat, we tend to see a reduction in genetic influence—that is, the predictive power of the PGI. When students were forced to stay in school longer in the UK—that is, the age at which students could legally drop out of school was raised—we see that the effect of the education PGI dropped on both education and on BMI. This British schooling reform that forced people to stay in school made the educational system there a bit more like that of Communist Estonia—where individual choice was subjugated to the collective goals of the state. The result: more equality and less influence of genetics. An interesting combination to ponder with respect to what meritocracy means. Indeed, if we all raised children from embryos on kibbutzim, giving them the exact same environments, we would see *only* genetic differences since environmental factors would be eliminated. The difference is that in my mythical kibbutz, choice for the children would not be constrained; only the systematic effects of environmental differences that they experienced by accident of birth would—wealth, social status of their parents, and so on. In this progressive, lenient kibbutz, if one kid wanted to spend more time doing math and another wanted to play soccer, they would be allowed to do so. Their PGIs would express themselves fully, uninhibited by externally imposed environmental constraints. In the above examples, we are limiting the expression of genotypes by forcing a thousand different PGIs onto the same path— for example, staying in school.

Within these different environmental contexts—opportunity, information, and choice versus constraint—is it better to be an orchid or a dandelion? Well, that depends . . . on the environment. If you knew that you were going to receive the exact nurture you needed and that no challenging environmental shifts were on the horizon, you might opt to be an orchid. But if you weren't sure where you were going to end up sprouting, and under what conditions, then being born as a dandelion would be the better bet. If you knew that you would not experience the kind of neighborhood I grew up in, having an emotionally sensitive genotype might be an advantage—making you more empathetic, sociable, and so on. But if you grow up in a violent area afflicted by an epidemic of addiction, maybe it's better to be genetically stoic, so to speak.

These are two different evolutionary strategies. Being an orchid pays off when the environment is just right—say when there is rain-fall at least once a week. But you wouldn't want your population to consist of all orchids, because one year there may be a drought. That year all the orchids, who had been dominating the population with their reproductive fitness, die out. Only the dandelions, who can survive any sort of weather, make it through the dry spell to keep their species going. So ideally, you would want a mix of orchids that take maximal advantage of conditions when they are ripe, and some dandelions that will survive no matter what nature throws at them. This is called bet hedging. What the exact ideal mix is depends on meteorological variability.

I'm talking about evolutionary strategies here, but it's important to keep in mind that for most human outcomes—education, personality type, BMI, and many others—we are not so much talking about the orchid-dandelion distinction in terms of survival and reproduction (the name of the game for evolution), as we are in the case of these flowers, but instead, we're simply using this evolutionary metaphor in terms of strategies to thrive in our present society. That's not to say that gene-environment interactions don't matter at all for human survival and evolution. But we have to keep in mind that from an evolutionary point of view, the only factors that matter for natural selection are 1) Do you live long enough to reproduce? 2) How many children do you bear? 3) How fertile are those kids? If smoking only kills people after they pass on their genes to the next generation (which is generally the case) and doesn't affect fertility (though nicotine can lower sperm counts, effects of tobacco on overall fertility are small), the genes that are pro-smoking will continue to survive in the population. Likewise, if less educated people, short people, or people with genetic tendencies toward depression have as many kids as those on the opposite ends of these genetic spectra, then the respective PGIs are not selected for or against in the population. Moreover, how these PGIs differentially interact with a variable environment—that is, GxE—doesn't usually affect if and how those genes get passed on. That is, the gene-environment interactions effects we have been talking about—like Vietnam and smoking, for instance—are probably not playing a role in human evolution. That doesn't mean they aren't super important for predicting individual experiences.

That said, in some more extreme cases, gene-environment interactions may actually play a role in determining whether our genes get passed on. I'm thinking here of the two sisters from my neighborhood: Leila will likely never have children at this point in her life. Her genetic road ends with her. This is the case for most agoraphobics. Crystal, meanwhile, has two children. In this case, the sisters' respective genotypes may have indeed had an "evolutionary" effect, with Crystal's genes surviving and multiplying in the population. Whether Crystal is a dandelion and Leila an orchid would depend on whether in some alternate universe where Leila lived in a stable, calm, and supportive community, she might have borne *more* than two children. As much as I'd love to get some saliva samples and run PGIs for both of them to see whether their genetic "prisms" were different, without fifty Leilas and fifty Crystals randomly assigned to rough and calm neighborhoods, I'll never know if my genetic intuitions were correct.

Understanding where and when GxE interactions are at play in the world will make who thrives and who falters seem a lot less random. Right now, if I ran a careful experiment that tested a new reading curriculum in grade schools and found that my intervention did, in fact, raise reading scores by 10 percent in the treated group as compared to the control group, that would be very exciting. I could probably get that result published in a high-profile journal; it would get some press attention; and, maybe, it would eventually affect how we teach reading in public schools. But obscured in that plus 10 percent *average* treatment effect may be the fact that for half the kids, there was no effect at all. For another quarter, my new approach to teaching actually had a negative effect. But the last quarter of kids doubled their scores. When we put all these heterogenous responses together, we get my average 10 percent benefit. But that's not fair, especially to the kids who were hurt by my new approach. What if we had our genetic prism and we could actually see which kids reacted differently by genotype to, say, some sort of "phonics" versus "whole language" PGI?[27] Suddenly, we are in a world where we are not all subject to averages, where biological diversity is not ignored, and where we can get a handle on why people like Leila and Crystal or the kids in my experiment react so differently. Armed with this information, we can design a better society.

This mutual dependence of genes and environment on each other for their effects augurs a world with genetically tailored environments in the domain of healthcare, education, and so on. Once we know what environmental conditions or interventions work or don't work for specific PGIs, we might want to act on that information, with all the necessary caveats about the efficacy of the PGI, privacy, stigma, fairness, and so on. For example, if we know that a PGI for psychopathology predicts which veterans are more likely to suffer from PTSD (even, or especially, if they would have suffered no mental illness absent their exposure to combat), we could direct those members of the military with risky PGIs toward military occupations with less risk of such psychological trauma. In school systems, we could design individualized education plans based on genetic risks and strengths once we realize how particular pedagogies interact with genotypes. Or, we might want to identify underserved poor kids with high PGIs for educational attainment to recruit them for magnet schools with advanced courses so that their environments no longer hold back their genetic potential. Who knows what such a controversial policy might do to overall inequality in the long run, but it would certainly help those underserved children who got recruited in the immediate term.

Finally, policymakers from the Congressional Budget Office to urban planners to public health officials might even take genotype (PGIs) into account when they model how humans will react to societal changes. Recognizing the importance of GxE interactions is simply acknowledging that humans differ from each other and will react to environments imposed on them in different ways, but explicitly modeling GxE effects might allow us to maximize the social welfare of those different "types" of humans with differentiated environments rather than imposing a one-size-fits-all approach that will be good for some and bad for others. (Though, when compared to more universal approaches, such a policy framework may impose other, political costs in an increasingly polarized society.) We are, of course, a long way off from this idealized version of genetically informed policy since the current crop of PGIs tend to be noisy predictors and only work for one subpopulation (those of exclusively European descent).

More importantly, by revealing important GxE effects on top of showing how nature and nurture constitute a false dichotomy in other ways, sociogenomics should end the bitter war between the hereditarians and the blank-slaters. Yes, we humans are animals with brains and bodies, each constructed differently based on our unique DNA blueprint.[28] But when those "specs," those DNA-based plans, hit the social and physical outer world through our bodies interacting with it, all the differences in what they are able to experience and what feedback they get affect the ultimate pathways those bodies take. Perhaps the best analogy is not a simple prism but a series of mirrors and prisms, bending and bouncing colored lights off of a disco ball to create the complex light show we call human society.

8

Two Lotteries

We tend to think of nature as happening inside our bodies: How many fast-twitch muscle fibers did we inherit? Are we short or tall? Do we metabolize alcohol efficiently? These—and many more—bodily attributes affect multiple aspects of our lives—not just health and wellbeing but whether we marry and reproduce, how far we go in school, and what position we occupy in the hierarchy of society. Nobody questions the notion that outcomes like height and breast cancer, eye color and schizophrenia are deeply influenced by the genes we inherit and that, in turn, affect who we become.

Meanwhile, when we tally up the ledger for the environment, we typically think of accidents of birth, or what I call the "shit happens" theory of the environment: Did we experience a devastating car accident in our teen years that caused us to miss a year of school and walk with a limp for the rest of our life? Were we born into a society that discriminates against a group to which we belong? Did we grow up in poverty because our dad died of brain cancer when we were a toddler? Sometimes what happens is not shit but rainbows and unicorns instead. Did the childless widow next door leave everything to us in her will? Did we randomly get assigned a particularly inspiring eighth grade math teacher?

The examples I've just provided of both the influence of genes and that of the environment were carefully curated. On the nature side, I picked attributes where the effects of genes do not depend on social interaction to come to fruition. While height, for one, is to a minor extent influenced by our environment—namely, whether we get enough nutrition or take certain medications that might stunt our growth—there is no way you are going to produce a strapping six-foot-two body out of genes that have a growth algorithm that aims for five-foot-four. And absent major deprivation, the body with six-foot-two genes will never end up five-foot-nothing. The height genes don't depend on a series of life choices and the seizing of opportunities by the individual possessing them in order to come to fruition. What's more, the heights of the people around us do not affect our own height.

Likewise, on the nurture side, all the examples I've provided have nothing to do with our own actions or the genes inside us. They do not depend on us navigating ourselves into certain environments. If two sisters happen to come of age in Ukraine in 2024 during the Russian war on that country, neither is ever likely to become a world-class violinist because growing up in a war zone means that they won't have access to the necessary instruments or training at the critical window of development. Moreover, they have more important things to worry about—like surviving snipers and mortars and getting enough food and water. Critically, this extreme environment has nothing to do with their genes. The war affects the sister with perfect pitch and rhythm as well as the sister who could never carry a tune.

Whether they are aware of it or not, when most people think of the genes-environment debate, they are posing these two extremes against each other. But as we have seen, recent discoveries in sociogenomics show that this division is much too simple and, in fact, is more the exception than the rule. Indeed, sociogenomics describes a much more interesting world where the boundary between genes and environment is illusory. Perhaps the best way to illustrate this new reality is by reexamining the example of musical talent. Let's imagine a less extreme environment than a war zone and talk instead about a wealthy family of four daughters growing up in the U.S. today: Sandra, Sue, Sarah, and Sadie.

Sandra, the eldest, has inherited a certain architecture of her inner ear optimized for identifying musical pitch, along with a high level of fine motor skills in her fingers. Having been dealt less musically fortunate genetic cards, Sue, the second born, is clumsy when it comes to using her hands, has little lung capacity, and can't recognize harmony from dissonance. When the parents of these privileged girls sign them both up for music lessons after school, Sandra enjoys the experience, and Sue does not. Indeed, unless she is a glutton for punishment, Sue takes the first opportunity to give up on musical endeavors because it is simply not fun to keep doing something you stink at.

Sandra, on the other hand, is hitting all the right notes—not to mention receiving praise from her instructors. This positive feedback may cause her to ask her parents to send her to music camp that summer. The sisters' respective genes, then, have guided them to distinct environments. If she does well at that camp, Sandra might want to audition for a music conservatory rather than attend a regular high school. She is finding her niche—but if she doesn't continue to expose herself to more and more enriching environments, her musical talent might wither on the vine.

Sandra does persist, though, immersing herself in the right environments to hone her genetic abilities (i.e., active GE). Part of the reason she does is because she has the physical attributes that make musical experiences rewarding. Another part of her persistence may simply be that she has genes that cause her to seek new experiences and persist when she meets challenges (epistasis). (I am stipulating that there are such genes, and there is indeed growing evidence for my claim.) That sets her apart from the third sister in this family, Sarah. Thanks to the luck of the genetic draw, Sarah was born with the same ear and manual dexterity as Sandra but simply has less innate (read: genetically influenced) drive in general and thus does not push herself to audition for the conservatory. She is not Type-A; she's happy just playing tunes for her friends and family. The point here is that part of the genetic effect that leads Sandra to the heights of the music industry works through the environments she *chooses* to expose herself to (and is able to immerse herself in, absent bias, famine, and so on). Only Sandra clamors for more piano lessons. In Sue's case, even if she had asked for more lessons (or her parents had imposed them on her against her will), it would not have made a difference on account of her tin ear. But in Sarah's case, the raw musical material is there: it is her *other* genes related to personality and ambition that prevent her from pursuing the environments that would mold her other genetic inputs into those of a world-class musician.

The bottom line is that for most traits we care about, genes express themselves by causing people to alter their environments or to seek out particular environments—social environments, to be more specific. If we block off those critical social environments—thanks to war, poverty, discrimination, or other factors—the genes don't take their effects (a gene-environment interaction). Aside from these factors outside our control, our genes choose the environments to which we are exposed based on our innate preferences and the social feedback we receive. In other words, a good part of the environment is not just shit that happens but shit we *make* happen—or rather, that our genes make us make happen.

But wait, we haven't discussed the fourth sister: Sadie. Sadie was born with the same innate physical potential for musical talent that Sandra enjoys, as well as the same genes for grit, motivation, and persistence. Unlike Sandra, who is considered conventionally attractive,

Sadie is not considered pretty by most people in the contemporary U.S. Her parents and instructors don't enjoy watching her perform as much, although they are probably not even aware that their lack of interest is due to Sadie's appearance. What's more, when they hear her play, they don't think she sounds as good as her sister. As a result, they don't encourage her as much in her musical pursuits, and Sadie does not end up going to conservatory despite having the same melodious genotype and ambitions (i.e., genetic signature) as her sister. It could be that Sadie asks to audition and her parents or teachers say no. Or a much more subtle dynamic could be at work in which Sadie, like many other less attractive individuals, is more likely to internalize the idea that she is simply no good at music. Either way, the result is the same—because of appearance-based social discrimination, genetic talent for music does not get realized. The debate becomes more nuanced still: Is this sort of social discrimination genetic or environmental? Technically, since beauty is largely influenced by genetics, it's an effect of Sadie's DNA, but how others treat her stems from the social response that some of her genes evoke. But the suppression of her musical ambitions is based on social bias rather than any strict linkage between the biological effects of her genes and musical talent. Let's call this *genetic shit happens* (aka evocative GE). What kind of genetic shit happens depends critically on the kind of society we live in; namely, there is nothing biologically universal about such effects. In a community where discrimination against red heads is rife, the genes for red hair will lead to negative outcomes. In our society, meanwhile, the genes for skin tone unfortunately lead to all sorts of environment treatments that have no rational relationship to the actual functions of those genes.

To further blur the line between genes and the environment (in what I think is the most mind-blowing way of all), let me introduce you to another family in addition to the four daughters we just met. This family has a son and a daughter. Jennifer enjoys the same fortunate genetic signature when it comes to musical talent, grit, ambition, and appearance as Sandra. Sandra and Jennifer met at conservatory, in fact, where they discovered that they both have the same driven personality type that has gotten them this far.

But Sandra arrived at that golden genetic ticket by drawing a straight flush from her parents' genomes. When it comes to the genes that influence musical talent (or drive), her parents each had a copy of the "melodious" version of the gene and a copy of the "tin ear" version of that gene. They, themselves, weren't musicians, so they didn't seek out a partner with the best sense of rhythm or ability to carry a tune. They were both average in this regard and found themselves attracted to each other on other dimensions (say, their mutual love for literature). But even by not actively choosing each other on the basis of musical talent, they ended up similar on this genetic dimension because other people who were on the talented end of the musical spectrum picked each other, leaving the average people, like Sandra's parents, in the marriageable pool. From a mix of these relevant genes, Sandra happened to get the "melodious" gene from each parent, giving her the double dose of pro-music genes necessary to be competitive on the world stage.

Meanwhile, Jennifer sprung from the loins of parents who both already had those double doses—her mother and her father each had two copies of the "melodious" genetic variants. So, it was a fait accompli, genetic predestination, if you will, that Jennifer would have this gifted musical genotype. This, too, was due to genetic sorting, since her parents matched on their love of music.

Imagining that other, random aspects of their respective environments—like family income, race, and so on—are the same, would Sandra and Jennifer have an equal shot at a successful music career once they get to conservatory? It turns out that even though they have the same double-dose genotypes, Jennifer enjoys an edge in the world of music. That's because the genes of the girls' parents—

even the ones they don't pass on—shape their environments in measurable and important ways (genetic nurture). Jennifer's nameless brother is, by definition, genetically like her in terms of raw musical talent since the only genes any child of their parents can inherit are the melodious ones. Sandra's sisters, however, are likely to be musically average. (Indeed, she has two with good musical genes and one with poor ones.) In other words, while Jennifer's childhood home— perhaps the most important environment one experiences in life—is filled with musical interest and talent, Sandra's is not as likely to be. Because of that fact, Sandra is not as likely to succeed at conservatory since her environmental experience gets shaped by genetically programmed family, friends, and more.

Even when Jennifer and Sandra both make it to conservatory, the distribution of genes among their peers will also affect the environment in which their talents marinate. If Sandra happens to get a roommate at conservatory who is serious and driven, yet generous with her support, she might thrive, while Jennifer might wither (despite her family's musical pedigree) if she ends up bunking with a student who is spiteful and undermines her at every turn. Peers with musical talent will push them both to hone their talents to an extent that would not have occurred had they not been exposed to such colleagues. The same is true for the quality of their instructors. The point here is that a part of the social environment that affects us is made up of the genes of others. We call these effects *social genetic effects* or the *social genome.* Social environments are, in part, genetics one-degree removed.

Finally, how we react to environments depends on our genome. Or, put conversely, how our genes matter depends on the environments we happen to encounter—that is, gene-environment interaction. Even when we are exposed to random environments that are given in equal "doses" to us irrespective of our genes, the downstream effects of those environments depend on what's in our DNA. Conversely, the effect of genes depends wholly on the random environments in which we find ourselves. To return to an earlier example, it is likely that the genes my neighbor Leila inherited by chance (and her sister did not) were the reason that when she was exposed to a violent neighborhood, she became agoraphobic. Absent the tough environment, the two sisters might have both become scientists.

These complications to the simple nature-nurture dichotomy are changing our understanding of what makes humans who they are. To recap: our genes guide us through and even evoke the environments that shape who we become—that is, genes don't "work" fully without the "nurture" they clamor for and engender. Take verbal ability. There's no doubt that being read to a lot when one is young is an excellent predictor of scores on the verbal section of the SAT a decade and a half later. But it is not random which kids get that kind of attention. Nor is it very surprising. Kids who have a genetic tendency toward high verbal ability are the ones who beg their parents for "just one more story" before bedtime. As they improve their reading, they get further environmental investments from schools and teachers who may put a kid who loves reading into an advanced track or gifted program. But those investments on the part of parents and society are given to the select few with innate promise. In other words, the environment is crucial, but our choice of ponds—the music conservatory, the athletic field, or the computer lab—is often a critical pathway by which our genetic differences express themselves via the proverbial ten thousand hours that are needed to master a skill.

Second, those social environments, in turn, are made up of a social genomic soup of the DNA of others. The social world we encounter in our random (or not so random) walk through life is made up of people whose actions toward us are conditioned, in large part, by their genes. Part of the social environment, then, is just genetics one degree removed. Third, what social genome we "happen" to experience is not random—it's the result of genetic sorting at all levels of society, from who we befriend or marry to what state we choose to live in (or our parents chose for us).

And, lastly, how our genes matter depends on the environments to which we are exposed. (And vice versa—how environments influence us depends on our genes.) Being zaftig could make us the hottest commodity on the dating market or lead to many Saturday evenings spent alone depending on the social norms of a given era. Having an aggressive genotype could land you in the C-suite if you were born to a family with the resources to provide opportunities in the business world to its offspring, but it could land you in prison if you happen to grow up in a crime-ridden community.

I WENT INTO SOCIAL SCIENCE WANTING TO KNOW HOW DIFFERENCES in life conditions—which I saw in such stark relief during my childhood spent commuting across the socioeconomic landscape of Manhattan (the most unequal county in the U.S.)—affected our chances to succeed in contemporary, capitalist society. In this vein, my intellectual journey started with an investigation into the impact of parental wealth in explaining race and class differences in socioeconomic attainment. But almost from the get-go, I was troubled by the fact that the lottery of birth was, in fact, two lotteries: one that determined your childhood environment and another that determined your genetic makeup.

These two lotteries overlap and are intertwined, so there was little I could do to figure out what really mattered other than to find little, accidental experiments that varied specific environmental conditions—from prenatal nutrition to wartime military service to the gender of one's kids—and assess the impact of each of those factors in isolation of genetics. But I never found a great natural experiment for the core issues I cared about: race, wealth, and parental education. After wandering in the desert of the economics credibility revolution for a good decade, I found genomics. Now I can directly factor out much of the genetics that was leading folks to self-select into environments, thus revealing a "purer" (i.e., more accurate) sense of how the environment matters—even if I can only do that within groups and cannot explain differences such as race and gender gaps. (I still need sociology for that.) But that time in the desert was not for naught. The approaches I learned from economics turned out to be critical to understanding how the random environments we find ourselves in work together with the random genomes we inherit to produce unique outcomes—that is, GxE interaction. It's still early days, but I feel that finally social scientists have the pieces of the puzzle to put together a more complete picture of human variation.

Of course, we didn't have this technology when I was growing up on New York's Lower East Side. But since our DNA doesn't change, I could theoretically go back today and collect saliva samples from my childhood neighbors (at least those still living) and calculate PGIs for each of them (ignoring the fact that PGIs don't translate across ancestral groups for this thought experiment). With this information, I might be able to make better sense of who faltered and who made it. While, of course, PGIs don't tell us with much accuracy whether a particular individual will graduate from high school or suffer from major depression, if I am interested in explaining, on the whole, why one tranche of my neighbors suffered from mental distress thanks to the conditions back then while another slice of the community seemed relatively unperturbed by the violence and stress that permeated the streets, PGIs might serve as a useful prism to expose those patterns.

But that information might do more than just explain who made it out alive and well and who didn't; it might actually help some of those in the latter group. If, for example, Leila's mother had known that she had a sensitive genotype with respect to psychopathology, she might have done whatever it took to get out of town, so to speak. Ditto for parents of children with high PGIs for substance addictions.

Likewise, perhaps if school officials had known about kids' genetic propensities for dyslexia, ADHD, and so on, they might have tailored educational plans to better serve their students' individual needs. Those with promising educational PGIs for math, verbal ability, or other talents might have received extra opportunities to realize the promises of those genotypes.

Armed with information, we can take action. How we act on that complete picture is the big question. Should we aim for genetically customized policy? Should we use PGIs to predict outcomes and then act on those predictions through college admissions decisions, insurance prices, and so on? Should we provide free genotyping to everyone, in order to democratize this information revolution? Earlier in my academic career, I was constantly offering policy recommendations based on my research. Once I started working on PGIs, however, I became trepidatious. I see the pros and cons of, for instance, banning insurance companies from risk-adjusting based on customers' PGIs. I see my role now as a nonpartisan advisor to a legislative committee rather than an aide to the legislators themselves. I lay out all the pros and cons and encourage a robust public debate about our options in this brave new world. One thing's for sure, though: we don't want to enshrine a version of genetically informed policy that embodies the eugenics view, only redoubling genetic differences with social (dis)advantages by, say, showering resources only on those who appear—under the current arrangements of society—to enjoy genetic privileges in a given domain, especially since some of those genetic edges may be due to irrational factors like height or skin tone.

Taking into account a complete picture of human variation—that is, the combined genetic-environmental influences on outcomes—in public policy decisions may make people uncomfortable, but ignoring or denying genetic inequality isn't going to make it go away, and it isn't going to mean that our policies won't impact people with different genetic makeups in different ways, creating inequalities in the process.

It's important to recall that genetic inequality is widening, whether we like it or not. If we do not consider the impact of genes, socioeconomic and genetic polarization will keep accelerating. With the knowledge we've gained from sociogenomics, however, we can now see that the impact of genetic inequality is not inevitable but rather a function, in part, of what choices we make as a society. Now that PGIs are providing a more complete picture of nature and nurture, we can use this knowledge to help people like Leila and also improve society overall.

The good news is that genetic inequality *is not* intractable. Because nature and nurture are part of the same Möbius strip, altering the environment can affect how our genes work, thereby ameliorating genetic inequality and the social problems it causes. Not only are heritability estimates and current PGIs population-specific, but they are also environmentally specific. Alter the environmental landscape and genetics can gain or wane in importance. Recall that the education PGI predicted outcomes well in capitalist Estonia but not in communist Estonia. The social structure of capitalist schooling drove the heritability up. What we know about the predictability of PGIs or the total heritability of traits is only for the particular society and the particular moment at which they are being studied. Introduce radically different policies—universal basic income, mandatory higher education, free early childhood education for all, and so on—and the one thing we can be sure of is that the predictive power of our current PGIs will change.

It's true that there is precious little we can do to raise the adult height (a trait almost entirely linked to genetic differences) of the least tall among us, notwithstanding dangerous bone lengthening surgeries. But it is also true that myopia is highly influenced by our genes and yet, there is a very simple, inexpensive intervention that will fix nearsightedness. You may be wearing them right now.[1] The result is that corrected vision is not heritable at all, thanks to glasses (and contact lenses and Lasik surgery). Or take PKU, the single-gene disease that prevents individuals from properly metabolizing the amino acid phenylalanine. On the ingredients list of many diet sodas, you may have noticed a warning that "this product contains phenylalanine." That's because the simple fix to this disease, discovered by British chemist Louis Woolf, is to avoid consuming products with that compound.

Likewise, the heritability of reading at age four in Australia is extremely high. But by age six, it has fallen dramatically. What's happened in the meantime? Kids go to public school, an institution that imposes a universal environment of reading instruction on them, causing the genetically driven differences in environmental feedback to matter less. If we imposed equality in our school experiences in this country, there is no doubt that both the racial and socioeconomic test score gaps would shrink; the heritability of cognitive ability would decline as well. Conversely, plenty of non-genetic conditions have no easy fixes: fetal alcohol syndrome, paralysis resulting from accidents, respiratory problems from particulate matter in the environment, communicable diseases, and so on.[2]

The first step in addressing genetic (or environmental) inequality is being able to measure it. PGIs help us better understand both genetic and environmental inequalities, thereby allowing us to affect them through our collective choices. PGIs will help us know the effects of schools, neighborhoods, income support, military combat, and so on for any outcome we care about by factoring out some of the relevant genetic influence to leave more purely environmental effects, unconfounded by GE correlation (the nonrandom distribution of genes across environments). Moreover, PGIs will also help us detect social segregation and polarization by providing a measure that is unchanged through experience, revealing only how we sort and not how we influence each other. In this way, the clustering of PGIs can help us diagnose geographic inequalities by political attitudes, economic prospects, and even mental health.[3] Seeing something clearly is the first step to being able to ameliorate it.

Whether or not we are talking about differences between groups or within groups, we must acknowledge that genetic differences for behavioral traits do exist. However, accepting this scientific fact doesn't mean becoming a reactionary eugenicist. The preceding examples show that we are not beholden to our genetic differences but can amplify or mitigate them in order to engender a thriving and just society. As sociogenomics descends from the ivory tower, we simply won't be able to ignore the importance of genetics or cling to the false promise of a blank-slate ideology. That's a sure way to cede the debate to reactionaries, who will be able to use the PGI to say that genes *do* matter, and that, therefore, some people are inherently superior to others. We will need to learn to hold a more nuanced line, insisting on the blur between nature and nurture and resisting the temptation of simple determinism.

"FOR YEARS, NEW YORK PARENTS HAVE BEEN APPLYING TO PRESCHOOLS even before their youngsters are born. That's not new, but the approach [of] one prestigious pre-school on the Upper West Side is [new]," an NPR story from 2012 reported. "At the Porsafillo Preschool Academy, all applicants must now submit a DNA analysis of their children." The admissions team, NPR reports, "is looking for genetic markers that indicate future excellence—things like intelligence, confidence and other leadership traits." The school faced pushback from upset parents of potential applicants. The leadership defended its new, biologically informed policy: "This is not unethical at all. If anything, it's extremely ethical. This is now no longer a subjective decision. This is a clinical test that can show us how a child will perform throughout its life."[4]

This story generated many angry calls to public radio stations across the country before people realized it was a joke, having aired on April 1. But before long, there will be a real story airing on NPR about a school screening applicants' DNA. We might also see individualized special (or gifted) education plans based on PGIs—in other words, genetic tracking. Progressives may push to use these tools to identify poor kids with high PGIs for education and shower opportunities on them.

The problem common to all these approaches is the fact that the education PGI captures all sorts of effects besides the ones we think of as legitimate—height, skin tone, and so on. These irrational effects will become even further baked into the social fabric of our society if we reify the PGI above all other judgement. Moreover, the PGI is always a noisy predictor, and other environmental factors matter a lot, too. It's very possible that someone with a high PGI for education won't thrive academically, or vice versa. So many other things come into play. Nor would we *want* to live in a society where the PGI was uber predictive.

In this vein, we don't want to risk creating a vicious circle whereby the PGI is used to slot people into roles in a mechanistic way, thereby becoming deterministic where it was once probabilistic in its impact. Namely, given that PGIs predict only a minority of the variation in most outcomes, one worries about a self-fulfilling prophesy whereby a PGI that predicts 25 percent of the variation in, say, education, ends up becoming deterministic in its results because we act as if it is fully predictive in the absence of other information. Along these lines, if education PGI scores ended up as the sole metric on which admissions officers decide who gets into college (or preschool), then they will go from predicting a quarter of the variation in college education to basically all of it. They will become a self-fulfilling genetic prophecy—genetic and social inequality would then totally overlap, not for any reasons of efficiency or equity, but merely by freezing our current social arrangements by enshrining them in the mythology of "natural" and universal effects of genes.

This kind of algorithmic prejudice is already all around us. Admissions officers, hiring managers, police officers, judges, people looking for dates, and advertising algorithms all make decisions based on limited information. When information on individuals is scarce, we rely on averages of groups. Even your FICO score is based on an average risk of default for people with a credit history like yours. These group judgements are how we make decisions, for better or worse. Employers shun applicants who check the box that acknowledges they've been convicted of a crime, even if a given individual may have extenuating circumstances. Such group judgements tend to fuel discrimination against minorities. An HR worker reviewing applications may be tempted to choose a white applicant rather than a Black applicant, given that Black people, on average, tend to receive worse educations than white people given the present state of U.S. society. PGI or no PGI, such judgements re-enshrine social discrimination. Along these lines, when faced with uncertainty, the PGI may become a eugenic, shorthand crutch for individual or institutional decision-making.

It's currently illegal to discriminate on the basis of genetic information in health insurance and in employment, but that leaves lots of other domains where PGIs can be deployed.[5] We already know about how the criminal justice system can use genetic information to solve cases: the DNA left at a crime scene can be linked back to a suspect. Even if a suspect has no DNA in a database, they can end up nabbed if their relatives do.

A more sinister scenario has people with high PGIs for antisocial behavior monitored or preventatively managed, a situation predicted by Phillip K. Dick's concept of "pre-crime." This is not happening yet, but just as PGI prenatal testing will soon collide with stem cell technology to change the landscape of reproduction, PGI-based prediction may combine with machine learning and AI to change predictive policing. Perhaps some enterprising criminologist will develop a PGI for recidivism. This number could then be fed into the algorithms that currently hoover data on defendants to recommend suitable sentences. A PGI that supposedly predicts recidivism may pick up all sorts of irrational factors such as hair type or facial morphology. Moreover, it would only predict in the context of our current—highly flawed—criminal justice regime. As a result, using the PGI to decide on sentences would reify the social inequality baked into that system.

Under faulty, neo-eugenic logic, at-risk youth with riskier genotypes might end up being monitored more closely than those whose teenage troublemaking results from unlucky environments under the misguided assumption that we can change the environment, but genes are forever. Conversely, a kid with a "good" PGI who strayed off the straight and narrow might be able to convince a judge to give them a slap on the wrist thanks to the evidence of his "genetic promise." Absent a nuanced understanding of sociogenomics and the Möbius strip aspect of gene-environment interplay, people may revert to the common genes-as-destiny cognitive shortcut and then make that a self-fulfilling prophesy. The fear is that even if PGIs can predict a tiny sliver of the variation in who commits crime, by treating those numbers as gospel, we may inadvertently create castes based on our genes. Instead of (or, rather, in addition to) racial profiling, we might live in a world of genetic profiling.[6]

PGIs could end up enshrining genetic differences in society in other ways by, for instance, changing our options as consumers. To price his auto insurance, Geico asked my then eighteen-year-old son sixty questions, yet they didn't offer a "discount" for sharing his personal 23andMe data. Today, they could easily add PGIs for education and for risky behaviors to their predictive models of accidents and save millions of dollars through risk adjustment. It may not seem fair to us that some people should get lower car insurance premiums than others simply based on what genes they happened to inherit, but the logic of profit maximization doesn't always care about fairness, in case you haven't noticed. The marketplace cares about efficiency, not equity.[7] Indeed, we have no problem forcing someone who has shown themself to be a bad driver to pay more in premiums. It seems fair. They are a proven "loser" as far as the insurance companies are concerned. But, in the same way, we don't tend to hold children as responsible for their actions as we do adults; should we not judge people on the basis of their genes, even if those genes predict important outcomes that affect us all (as does risky driving)?

Car insurance is a comparatively easy case since we are all required to buy it. When it comes to other forms of insurance that are not mandatory, one can see individual consumers deciding whether to buy long-term care or life insurance based on their PGIs for dementia or heart disease, respectively. If potential consumers of insurance are able to choose whether to buy it based on risk factors they know about but the insurance companies don't, this undermines the entire insurance market, to the detriment of everyone else. Only those with risky PGIs will want to buy insurance, and given how much it costs to insure these risky PGIs (a lot), the price of insurance rises. Those with less-risky PGIs will no longer want to buy insurance at the inflated price, so all that are left are risky buyers, and the market can die in a price spiral. Insurance is meant to be an equalizing force along a dimension of risk: those with high-risk or unlucky circumstances are helped by those who are fortunate enough never to need insurance in the end. The key is that none of us knows with any great confidence which camp we fall into—vis-à-vis genetically based inequality—otherwise, this implicit social contract falls apart.

Is the solution, then, to give genetic data to both consumer and company? That takes a step in the right direction but introduces a new problem: when we can all risk adjust (consumers and producers), then the whole point of insurance is abrogated, assuming our predictive accuracy is decent. Either only at-risk people buy it or, if producers are allowed to adjust premiums, then the price becomes so targeted to our risk as to not pool risk across buyers. Dental and vision insurance, because they are voluntary and bought mainly by people who know they will need dental and vision treatment, face this issue. An alternate solution is to make more and more types of insurance mandatory—like auto or health insurance—and then limit the ability of companies to risk adjust by PGIs—like the ACA mandate, but for life insurance, long-term care insurance, dental, vision, and so on.

Most of the discussion about potential abuses of genetic data in the past involve the identification of people—that is, invasions of privacy. This has indeed gotten scarier due to the ability to genotype smaller and smaller amounts of DNA.[8] But the risks of the PGI are different: the PGI allows institutions of all sorts to use your DNA to aid them in making decisions about you—admissions, premium rates, whether to show you specific dating profiles, and so on. We may end up in a brave new world of our own (and not the government's) making. As PGIs seep into the agora beyond insurance, what will that marketplace look like? Will we continue to pay to get our own genotypes only to offer them to companies the moment we are enticed by a discount? Our genome is a different kind of data than our credit card number. If we are defrauded through identity theft, we usually can be made whole, and we can certainly get new credit cards issued. Sharing your PGI with a company is more like sharing your Social Security number or your birth date: It's not something you can change. Moreover, while your SSN or birthdate is more or less random, the PGI is useful for much more than uniquely identifying you—it can predict your behavior.[9]

Any time you quantify something that was hitherto hidden or assessed in a less precise manner, you run the risk of introducing a sense of hierarchy—a horse race, if you will. In some dystopian future, I can imagine new, computerized editions of the *Racing Form*, ones that provide the background experiences (as captured by statistics like the FICO score) and breeding of humans (as represented by PGIs) in the same format that the current version provides past performances, workouts, and information about a horse's sire and dam. In this world, school admissions offices, people looking for a mate or for a sperm donor, and even insurance companies peruse this spreadsheet of sorts, each running its own algorithm for the data, handicapping us all for our chances of success or risk of default.

THERE IS A REASON GENETICS SHOW UP IN SCI-FI SO OFTEN—THE alteration of our biological makeup carries a host of ethical and philosophical questions. In Aldous Huxley's classic novel, *Brave New World*, genetic engineering is used to mass-produce children in five distinct castes. This genetically encoded social order defines each person's role in society. The dystopian World State that Huxley created makes us uncomfortable on multiple fronts. We rightly fear a world where our fates are wholly determined by our genes before we are even born. Moreover, we abhor a society that slots us into social roles whether we like it or not. It's dehumanizing. Similarly, allowing our selection of mates, offspring, or classmates to be subject to cold-eyed calculations about genetic maximization is similarly creepy.

John H. Evans, a sociologist and bioethicist, has argued that human germline gene editing (i.e., modifying the DNA of our offspring) effectively dehumanizes children, turning them from sui generis people into objects in a game of attainment.[10] When we can custom order something that's been given a rating—as if through Amazon—we see it as an object. In this way, gene editing our babies nudges them just a tad down a slippery slope of being perceived coldly in objectified, market terms.

Evans's argument also applies to babies who were not modified but instead selected. If you picked an embryo that was the highest on genetically predicted cognitive ability, and that child didn't turn out as smart as you had expected, would your love for them become slightly tainted? What if you maximized on one dimension—say, cognitive ability again—ignoring riskier PGIs for other attributes and that kid turned out to develop schizophrenia or major depression? We already blame ourselves enough as parents for the nurture we provide. Imagine how we might second-guess our choices if the very genetic essence of our offspring (and their descendants) lay in our hands as well.

Most early-twentieth-century dystopias, like Huxley's World State, focused on the power of the government. And for good reason. The twentieth century saw totalitarianism lead to mass murder, ethnic cleansing, and the repression of human freedom. Eugenics, for instance, was a project of the state. But, today, the biggest threat of eugenics comes not from governments trying to control our fertility but from ourselves and the logic of the free market. Just as it has turned out with the internet and with AI, in Western societies at least, the greatest threat to our autonomy in the twenty-first century comes not from a centralized government that controls our information or our genes. It comes from individuals and corporations pursuing competitive advantage in the marketplace. The results can be just as pernicious, however.

Given these concerns, in 2019, I thought it worthwhile to take an expansive look at what Americans actually thought about the use of polygenic indices. I led a team of researchers from Princeton University and the University of Chicago in conducting a nationally representative survey of nearly 1,500 Americans to assess public sentiment about the use of polygenic scores in medicine, education, dating, and beyond. Unsurprisingly, vast majorities of Americans—roughly 91 percent—declared that judging people on the basis of their genes was morally wrong (say, looking down on someone for a low-education

PGI). This was what we expected from prior, national surveys that included questions about genetics.

Thus, we were shocked that these same respondents found it acceptable to use genetic information to make predictions and decisions about individuals in a wide variety of domains. We had expected that genetically screening embryos would be the least socially acceptable application of polygenic prediction since it was both eerily sci-fi and flew in the face of many religious beliefs. (Sci-fi in the sense that even though it is already happening, most people don't know about it yet.) Americans, however, tended to be fine with that practice.

Generally, cases where polygenic prediction empowered the genetic "consumer" to make a "better" choice—which sperm donor (83 percent), which suitor (79 percent), or which embryo to choose (68 percent)—were considered most acceptable. In cases where a faceless institution—like a school or an insurance company—was using your DNA to pass judgement on you, there was less public support. For example, people were the least happy about PGI being used for insurance pricing; only 38 percent found that use acceptable.[11] These findings held true across political affiliation and, remarkably—given the history of pernicious discourse about genetic "inferiority"—across race.[12]

If U.S. respondents had been universally against the use of PGIs across these various domains, the policy implications might have been obvious: we should simply ban their use. However, given the apparently high degree of public acceptability around the use of PGIs, the policy choices surrounding this new technology are not straightforward. To wit: Americans seem primed not only to permit—but actually *prefer*—that embryos created through in vitro fertilization are screened for polygenic tendencies. Such a practice could lower the disease rate in society in generations hence as screening for Down's syndrome does now.[13] On the other hand, such a practice of prescreening embryos may push us even further toward a caste-like society in which our high level of socioeconomic inequality is baked into the genes of the next generation. That's because wealthy parents would be more able to avail themselves of genetic-prediction technology than poor families—unless, of course, we chose to publicly fund such embryo selection for everyone.[14]

Policymakers and experts will need to get involved in the conversation about genetic prediction; the process should include not only congressional hearings but guidance from the National Academy of Sciences—whose report on gene-editing CRISPR technology, which recommended a worldwide moratorium on human germline gene editing (i.e., gene alterations that are passed on to the next generation), could be a model.

The good news is the fact that we found no discernable partisan differences on attitudes means that there is a window now open for constructive debate and decision-making. I suspect that the only reason that PGI usage is not yet politicized is because people aren't yet talking about it much. That will change the moment an insurance company asks you to spit in a cup or when a couple decides to abort their low-PGI pregnancy. Once PGI becomes a household word, shouting will surely replace dialogue. After all, masks weren't all that political either in 2019.

What would such policies in a world awash in genetic information look like? They might follow a framework whereby we actively work to mitigate genetic inequalities, to help people who are less fortunate in the lottery of conception.[15] In the domain of health, it's easy to see the promise of PGIs for achieving this goal. Screening and intervening early for any number of conditions ranging from heart disease to hypertension to addiction, for example. When conditions cluster—like anxiety, addiction, and depression—PGIs might help us sort out which is the root condition and which follow as sequelae from that primary diagnosis. This, in turn, could lead to more effective treatments. This logic of early, preventative diagnosis of risk might be extended beyond health to domains such as education.

By way of example, currently, Title I of the Elementary and Secondary Education Act of 1965 provides additional funding for schools based on the number of students they enroll who are from poor families or neighborhoods. What if, in addition to these funds, the government also provided additional monies that followed pupils with low PGIs for education in order to provide extra resources?[16] A controversial idea, for sure, but one worth debating, if only to reject.

PGIs might also be relevant to section 504 of the U.S. Rehabilitation Act that governs special education. PGIs might, for instance, be able to inform us which students are likely to develop special needs long before symptoms get diagnosed. We could intervene with dyslexic-potential children to teach them to read before they have fallen behind their peers. ADHD-prone students might be provided with cognitive behavioral therapy before they become disruptive. Ditto for children with high scores on the autism PGI. While there may be some squeamishness about diagnosing (i.e., labeling, or perhaps stigmatizing) children absent actual manifestation of conditions, if we are talking about added resources—rather than punitive measures—the potential benefits might outweigh the risks. Maybe.

PGIs that predict our genetic aptitude might also help us figure out what our talents are. This would serve both individuals and employers. Again, as long as PGIs are seen as value-added information and not wholly determinative (which they are not), then the positives may outweigh the potential harms done by incorporating genetic information into our decision processes.

There are nuances to a genetically informed policy heuristic, however. For instance, when we provide new opportunities to a large swath of society, as in the case of opening up higher education to women in the latter half of the twentieth century, we find that such opportunities are disproportionately taken advantage of by those who have certain PGIs, thus redoubling genetic inequality. By contrast, when the UK made an additional year of high school mandatory, research described earlier showed that those who are most genetically disadvantaged (and who would have likely dropped out earlier if they weren't required to stay in by the reform) benefited the most. We can see the same dynamics with respect to health outcomes. When the caloric environment was one characterized by scarcity, we didn't see high variation in body mass index (and genetic influences were repressed). When we made calories widely available, overall inequality in weight increased, and the genetic influences increased as well.

These patterns suggest that we may not need to know everyone's individual PGI to reduce genetic inequality. We can follow a simple formula: whenever policy provides opportunity to take advantage of new resources or information (i.e., we have more freedom to choose), it will exacerbate overall inequality *and* genetic inequality. So, if our goal is equality, we just need universal and mandatory policies.[17] In this vein, an "anti-eugenic" smoking policy might not be to tax cigarettes as we do now, it would be to pass a societal ban on smoking full-stop (as unrealistic as that might be), since existing policies seeking to make it harder to smoke have also created more, not less, genetic stratification in this health behavior and its downstream consequences, even as they have been effective at lowering the overall rate of smoking in the population. But, of course, there are major costs to individual freedom and autonomy when we pursue blanket bans such as this—we all know how popular Prohibition was.

Of course, like nuclear fusion, research on viruses, AI, and any number of other technologies, the PGI is itself a neutral technology that can be harnessed for the betterment of society or that can be abused for extractive profit, group oppression, or worse. However, what constitutes "betterment" is a complicated question: What if a genetically tailored policy helps some people a lot but hurts others a slight amount? What if it makes everyone better off, but those who were already advantaged gain the most? Such concerns are not unique to PGI-based policy choices but have bedeviled political debates for as long as humans have formed societies.

WHEN I TEACH SOCIAL STRATIFICATION—A BREAD-AND-BUTTER course in sociology—the students typically come in with very strong feelings about what's unjust in society, notwithstanding their presence at an elite university that enjoys the largest per capita endowment in the world. They feel that racism is rampant, along with other forms of discrimination. They think that inequality is simply the result of rapacious avarice on the part of corporations and the rich people who own them. (Sociology majors tend to be a self-selected ultra-liberal bunch.) But even with these views, almost nobody these days seriously argues for communism. They might want higher taxes on the wealthy and to redistribute these funds to the poor in the form of universal basic income, but precious few people are saying we should abolish private property or enforce strict economic equality. Many of them believe, deep down, in the ultimate fairness of meritocracy—the idea that the rewards someone receives should be commensurate with the results a person produces. (The students are perhaps a tad self-justifying in this respect.) Differences in results, they argue, come from differences in effort (and, to a lesser extent in their minds, talent). Notwithstanding my biased sample, this is not surprising; meritocracy has been the guiding ideology of the United States for decades, if not centuries. It is stitched into our collective consciousness as much as the idea of the U.S. as a melting pot—at least in mythology if not actual practice.

When I ask these students how similar parents' and children's incomes should be in a meritocracy, most answer that they should not be similar at all. To what family you are born should have no impact on what rung of society's ladder you end up on in a truly fair meritocracy. But then I remind them that even if we eliminated all differences in nurture by rearing IVF babies in an artificial womb and then raised them collectively on a kibbutz or randomly assigned them to adoptive parents such that nobody knew which child belonged to which bioparent, we would still see a likeness between bioparent and offspring. We have literally eliminated every possible environmental pathway in my science fiction-like scenario, but children would still, of course, inherit their DNA from parents. So, their likeness to their parents in terms of socioeconomic success would be whatever the genetic component of success was—perhaps 40 percent, perhaps 70 percent depending on our measure. This logic, of course, assumes we think environmental differences that children experience growing up are unfair and need to be eliminated in order to achieve the utopia of true meritocracy.

The students then quickly revise their answer upward from zero, getting my point that a meritocracy is a society that insists that children all grow up in equivalent environments, in which the parent-child likeness is only due to genetics, a society in which genes—and nothing about unfair environments—determine where we end up. Genetics, after all, measures our innate talents and proclivities for particular paths in life over a particular social terrain. I prod them more: Listening only to genetics in the way society sorts people is not only fair, I posit in Socratic fashion, but isn't it also the most efficient route to a world where everyone is working at their maximally productive capacity given the match between their role and their innate capacities? Pretty soon, I have most of them convinced—if still uncomfortable with the idea—that the more genetics matter and the less environment does, the fairer and more efficient a society will be. Viva genetic meritocracy!

Indeed, in my "utopia" of IVF-kibbutz babies, the heritability of traits would rise from what it currently is, because all family-based environmental differences would be moot. The only environmental effects that would manifest would be random events that befall or benefit some people and not others, thanks to chance. I tell the students that some scholars actually suggest we should use heritability as a measuring stick for meritocracy—evaluating the desirability of public policies based on whether they raise the heritability for, say, income.[18] According to this line of reasoning, a fair society is one in which heritability is close to 100 percent for any measure of socioeconomic success.

However, this would-be utopia, too, rests on hidden assumptions. First, we think it might be fairer because we assume that genetics picks up skills for the economy, that it predicts performance based on efficiency. Namely, if we strip down who makes it to the NBA or to the C-suite to genetics, that must mean that who gets there is who is best suited for those roles—winning basketball games or maximizing corporate profits. It ignores that genetic effects can be just as irrational—that is, mismatched to the task at hand—as environmental influences. Think back to all the irrationally evoked genetic effects related to beauty and the labor market, for instance. Second, it assumes that inequalities that result from the genetic lottery of conception are somehow acceptable—that is, fair—while those from the social lottery of birth are not. Environmental experiences are not necessarily unrelated to efficiency. Think of a random accident that affects someone's ability to do certain types of work.

Even if we wave away these concerns, a world of 100 percent heritability still sounds like some sci-fi nightmare in which everyone is stratified into genetic castes. I share the students' discomfort with so bluntly enshrining genetic determinism in society. The idea of your life chances being set at birth—or rather, at conception—with little hope of escaping your predestiny is the stuff that dystopian novels are made of. We didn't choose our genes, so such a society is not fair. Yet is it any fairer to allow unequal household environments—family background minus genes—to play a role in who wins and who loses? Or, for that matter, is it fair to allow random events—accidents, school placements, birth order, and so on—to determine who gets ahead?

Once we see the distribution of genes as the same as the distribution of environments—as a random accident of birth that has nothing to do with fairness—we hit a wall. We have to give up on the ideal of a meritocracy as a route to equality. This is not so easy to let go of for college kids who have scratched their way to the pinnacle of the system we currently call a meritocracy (even if they already recognize our present society is far from it). Rather than try to fine-tune our "meritocratic" system, then, perhaps the insights of sociogenomics should lead us to abandon it in favor of a different paradigm of what makes a "great society." Merit—who deserves society's fruits and who does not—may be the wrong frame altogether. That doesn't mean that we should abandon concerns with economic efficiency. Equal outcomes regardless of contributions to the collective good—a communist or fully socialist system—doesn't work either, for reasons that have filled up many volumes. But it may mean that we set an environmental floor below which we don't let anyone fall, regardless of their genes or environments. That's an old-fashioned idea called a safety net, but one that is only patchily practiced in the United States.

But maybe the fact that the line between nature and nurture is illusory will force a change in how we view fairness. The reductionistic view of genes and meritocracy sees genes as a causal agent somehow removed from the social context in which they are embedded. But seeing genetics as working through particular environments (over which we have some control as a society), and those environments being composed, in part, by the genes of the community (i.e., the social genome), means that taking genes into account as we think of how we want society to look and function is not so simple.

The discoveries in sociogenomics that reveal the Möbius strip relationship between nature and nurture should, one hopes, generate more public acceptance of people who end up in life situations different from us. In the same way that the discovery of biological (read: genetic) influences on sexual orientation took some of the moral sting out of "lifestyle choices," recognizing that genes influence important outcomes might make us more accepting of people who seem to be struggling with addiction, depression, or, in Leila's case, agoraphobia.

Furthermore, accepting that such life outcomes are not just predetermined results of genetic influence, but that they arise from the combined influence of DNA and the particular social arrangements we live under (not to mention our own genes, which form the social genome of others) would, ideally, increase not only our empathy for others, but our willingness to help them. This would be a rare and valuable (and perhaps improbable) combination of outcomes—an increase in empathy for those different from us, paired with the realization that we can actually *do something* about the inequality that emerges from those differences.

We could imagine a sociogenomically informed policy helping Leila by getting her out of our neighborhood. But society's imperative should be to improve the neighborhood conditions overall, for all the residents. The kids with particular learning needs might have been better identified and served; those with impulse control issues or antisocial behavioral tendencies might have been enrolled into programs to divert those energies to productive ends; and mental health services might have been provided to a lot of the residents to mitigate the trauma of the neighborhood violence. The key difference between sociogenomic and traditional policy is simply that the former considers how the genes of folks, not just their environments, matter in the evaluation of (dis)advantage as well as how policies might have different effects for different people based on their genetic makeup. If we give statins to people in their twenties who have very high cardiovascular disease PGIs—long before any sign of heart disease—might we consider proactive educational interventions for genetically at-risk youth? Might we assign desk jobs to PTSD-prone military personnel? At the current levels of predictive accuracy, the answers should be no; but in the future, these are real questions we will need to confront.

Just as it does at the level of society, sociogenomics teaches us as individuals that we have the power to act on or alter our genes' fates by changing the social environment. As people learn more about genetics in general and their PGIs in particular, they will hopefully see that genes are not determinative in the way that most people typically think of them in the context of single-gene diseases, and they may conceive of nature and nurture in a more nuanced way, changing our public discourse in fundamental ways. Just as people realizing they had gay relatives triggered changes in attitudes toward gay marriage and other related issues, when folks see that they have relatives who are predicted to be schizophrenic but who turn out completely healthy psychologically (or vice versa), they may change their whole framework for understanding genes. They will lament that they should have been six feet tall, genetically speaking, but only ended up at five-eleven—leading to intense speculation about what in their childhood environment caused the prediction error.

They will also learn that genetics don't drive outcomes on their own but depend on the active GE and niche-making that we ourselves do. A man who knows he has a high PGI for addiction might decide never to touch alcohol or other substances, standing that prediction on its head. By managing our own genetic tendencies, we will see genetic tendencies, as captured by PGIs, as deeply dependent on social context and our own navigation of the environments with which we are presented. We may fear that as the PGI leaks into society, genes will win Galton's declared war of nature versus nurture. But as we learn about our genetic information, it inherently becomes part of the social environment, reinforcing the Möbius-strip aspect of the relationship between the two factors.

What's more, the power of sociogenomics may allow us greater agency over our futures, not less. Our intuitions tell us that knowing that our behavior is determined by our genes is not compatible with a sense of personal agency. But as we learn more about our genetic risks and tendencies, they ironically become less fixed in their consequences, and we gain the power to act on them. Take the case of Angelina Jolie and the BRCA genes. When she learned that she carried copies of genes that put her at heightened risk for breast and ovarian cancer, she decided to have a preventative bilateral mastectomy along with an oophorectomy (removal of her ovaries to prevent ovarian cancer). Her genetic information became part of her informational environment, which in turn, affected her genetic risk for those two types of cancer—lowering it to essentially nothing—through her consequent surgical decisions. The bottom line is that the impact of her genes completely changed because of the social context. Moreover, her actions then became part of the information environment for other women. By going public with her decision—a risky move for a sex symbol movie star—she inspired other women to test for the gene and, if necessary, to take the same preventative actions she did. Indeed, in the wake of her public disclosure, the number of preventative surgeries spiked. Her BRCA test was not a PGI, but as people begin to act on their PGI results, the logic will be the same.

By undoing the categories of nature and nurture in these ways, sociogenomics will render a shift in how we think about cause and effect in our lives. Since cause and effect is, in essence, the fundamental narrative around which we organize the world, this is no less a cognitive readjustment than when our account of human events shifted from religious explanations for social arrangements (e.g., the divine right of kings) to secular ones; or when Copernicus displaced the Earth from the center of the heavens; or when Darwin showed that humans are different in degree, not in kind, from other animals. Sociogenomics represents the next such revolution.

Consciousness has been described in many ways. Some philoso-

phers talk about consciousness as the ability to integrate information.[19] Another idea, popularized in Douglas Hofstadter's famous
book, *Gödel, Escher, Bach*, is that consciousness is a loop. When we are
able to perceive ourselves, our bodies, and our thoughts, closing the
loop between our actions (motor control) and our perceptions (sensory
input), we become conscious. Key here is the learned link between
cause (I touch the flame with this object called my finger) and effect
(it hurts). At a certain level of consciousness, we no longer need this
motor-sensation loop since we perceive our own thoughts, recognizing the cause-effect loop without any physical actions whatsoever. ("I
thought about the sky because I thought about water, and they are
both blue.") Extending the loop metaphor: the more we know *why*
we think or act a certain way, the higher the consciousness we have
attained—hence the dual meaning of the word *consciousness*. Within
this broader paradigm, knowing more about the physical world, and
our genotypes in particular, elevates our consciousness to a higher
level by closing the loop between the cause of DNA and the effect of
who we have become.

Collectively peering into our genomes entails risk, but knowing how the stuff we are made of shapes our cells, our selves, and
our society is arguably the most important Enlightenment project of
all—more than quantum computing, space travel, or artificial intelligence. It closes the ultimate existential loop.

ACKNOWLEDGMENTS

Writing a book is the opposite of riding a bike: Ride a bike once, and you know how to for life. Each book, on the other hand, is a struggle to pedal forward, balance and steer, as if each manuscript were the first. Indeed, this effort felt *sui generis* in so many respects. Moreover, this is my first book in many years, so I definitely felt the enormity of the challenge. I was battling the tyranny of the blank page as if for the first time. I wrote an entire draft that I junked for parts, beginning again with a white screen. *The Social Genome* 1.0 and 2.0 both received copious and sage feedback (and other forms of support) from writers, editors, scholars, friends, and family members. Thus, I have many thanks to give—perhaps more than I have ever offered for anything I have ever written.

In no particular order, I want to express my gratitude to: Sydelle Kramer of the Susan Rabiner Literary Agency, for honing the book proposal with me for—no exaggeration—years. At Princeton, I am indebted to Josh Winn (of the Department of Astrophysical Sciences, for an "outside" read); Sam Trejo (for the most inside read possible); Tod Hamilton (for a careful read); Mitchell Duneier (for not only reading sections, but for being a generous chair who helped create the institutional conditions necessary for writing the book); and Matthew Desmond both for insightful comments he provided and for introducing me to Gareth Cook, President of the Verto Literary Group. Gareth and his colleagues Eli Mennerick, Pete Beatty, Kate Rodemann, and Siena Capone worked as my tough personal trainers all through the summer of 2023 and into the fall to crank out version 2.0 from the wreckage of 1.0.

The members of the Princeton Biosociology Lab over the years have been important interlocuters and collaborators. Some of those whose collaborative work appears in these pages include Ramina Sotoudeh of Yale University; Simone Zhang and Byungkyu Lee, both at New York University (NYU); and Fumiya Uchikoshi of Harvard. Other notable research collaborators mentioned herein include Emily Rauscher of Brown; Mark Siegal (my biology PhD advisor) of NYU; and Tom Laidley of the USDA.

Colleagues who gave critical feedback at various stages of the project include Ariane Conrad aka "The Book Doula"; Emma Daugherty (then a recent graduate of Princeton and now a member of the Norton family); and Florencia Torche (then at Stanford, but now my colleague once again at Princeton). Robert Darnell of Rockefeller University provided a close, "molecular" read. Jennifer Hochschild of the Harvard Government department offered a "political" one. Daphne Martschenko of Stanford and Emily Klancher Merchant of UC Davis both provided important critical reads from their perspectives as a bioethicist and historian of science, respectively.

The Robert Wood Johnson Foundation, which has supported my work in various incarnations over the course of my career (beginning with my postdoc), funded some of this research and provided a forum for discussing it in the form of a series of conversations at the foundation as well as a workshop devoted to what would become a central topic of the manuscript. Critical foundation staff include

Lori Melichar, my program officer, Nancy Barrand, Trene Hawkins, Jody Struve, and consultant Ben Milder of Burness Communications. The views expressed in the book are my own and do not necessarily reflect the views of RWJF or others. I am thankful to the workshop participants who, in addition to some of the people already named, included John Evans of UC San Diego, David Cesarini of NYU, and Molly Przeworski of Columbia. I am also grateful to the Russell Sage Foundation for funding some of this research both in the form of a grant as well as a stint as a visiting scholar.

At W. W. Norton, I am first and foremost indebted to Dan Gerstle, my editor, for taking a chance on a controversial topic by acquiring and nurturing this book. He read the manuscript multiple times, and his sharp eye strengthened both the argument and the prose. I am also grateful to Michael Moss, my textbook editor in the college division of W. W. Norton, who was also kind enough to provide comments on this book, keeping me true to my social scientific roots in the process. Zeba Arora offered helpful feedback as well as great copy for marketing.

Rebecca Rider provided insightful copyedits that doubled as a reeducation in English grammar—teaching me rules of usage that I didn't even know I didn't know.

Of course, I couldn't end without expressing my love and gratitude to my family. My kids, who keep sociogenomics real for me, from youngest to oldest: Tren Temim-Conley, Yo Jeremijenko-Conley, E Jeremijenko-Conley (a true blank-slater if there ever was one), and my stepdaughter, Jamba Jeremijenko-Rae. My mother, Ellen Conley, my sister Alexandra Conley, and my brother-in-law, Daniel Leonardi have always supported my projects. Most of all, I want to give a shout out to my beloved spouse, Tea Temim, who is the real scientist of the family and who is willing to boldly go wherever the data takes us.

NOTES

Chapter 1

1. Abraham Reichenberg et al., "Advancing Paternal Age and Autism," *Archives of General Psychiatry* 63, no. 9 (September 2006): 1026–32.
2. S. Sandin, "Autism Risk Associated with Parental Age and with Increasing Difference in Age between the Parents," *Molecular Psychiatry* 21, no. 5 (May 2016): 693–700.
3. At the time, the term of use was *polygenic score (PGS)*, but for consistency here, I am using the newer moniker, *polygenic index (PGI)*.
4. In 1969 the Berkeley psychologist Arthur Jensen published an article entitled, "How Much Can We Boost IQ and Achievement?" Jensen's answer to this rhetorical question was "not much" given the primacy of genes in his thinking—estimates had put the heritability of IQ at around 75 percent. Some readings of the article claimed that he was suggesting that racial differences in test scores were innate (though Jensen claimed this was a misinterpretation of his work). Nevertheless, he was roundly criticized, and protestors called for his resignation from UC Berkeley.
5. Some of these critiques were ill-formed. For instance, many argued that IQ and *g* (a common factor undergirding different types of intelligence) were worthless concepts. Some critiques of IQ dismissed the construct because a person's scores could change over time. But no serious scientist has claimed

that IQ is fixed at conception. There is plenty of room for environmental influences—like education—to affect scores. Indeed, estimated heritability peaks at 75–80 percent in middle adulthood.

Another critique is that the test-retest reliability varies with age. What this means is that if you give a passel of four-year-olds IQ tests on different days, the correlation between the scores on the two tests within each person is only 75 percent. But if you repeat that exercise with a bunch of thirty-year-olds, the correlation in performance across the two sittings will be above 90 percent. Anyone who has spent much time around four-year-olds doesn't need an explanation of why their test performance might vary dramatically depending on the day of the week. The fact that there's measurement error doesn't invalidate the underlying concept.

The deepest critique of IQ comes from Stephen Jay Gould's 1981 book, *The Mismeasure of Man*. Gould claimed that like phrenology of the nineteenth century, psychometric testing (and IQ in particular) was a pseudoscience meant to prove the superiority of whites. He showed how biased early tests were. He also claimed that combining different test measures into a single scale—*g*—was arbitrary and not necessarily indicative of a core, latent general intelligence. Moreover, calculating a heritability for that number, he claimed, was absurd. Such efforts, Gould argued, assumed that a correlation between family members in IQ meant that there was a causal relationship between their scores. He offered the metaphor of high correlations in over-time changes in his "age, the population of Mexico, the price of Swiss cheese, [his] pet turtle's weight, and the average distance between galaxies." Such a correlation, he concludes, does not mean that Gould's age rises *because* the population of Mexico is growing. However, this analogy itself represents a misunderstanding of how heritability is calculated and why it likely provides estimates of the causal impact of genetic and environmental variation on outcomes.

More recent work does question the idea that there is a common general factor, *g*, that is at the root of cognitive functions. Neuroimaging shows that different subtasks typically involved in a full-scale IQ test recruit distinct functional networks. Higher-order intelligence involves the recruitment of these different networks to solve a common task. This is taken to mean that general intelligence doesn't come "first" and then apply to different abilities (verbal reasoning, spatial logic, mathematical ability, and so on), but rather people have different abilities at these specific tasks, and general intelligence is a statistical construct—that bottom-up combination of different cognitive networks. That's not to say that people who perform well in one networked function don't also tend to perform well in others, just that general intelligence is more like an average or an overlapping set in a Venn diagram than it is a thing unto itself—say a diamond—which displays multiple facets through the subscales of an IQ test. See: Adam Hampshire, Roger R. Highfield, Beth L. Parkin, and Adrian M. Owen, "Fractionating Human Intelligence," *Neuron* 76, no. 6 (December 2012): 1225–37.

6. Murray acknowledges that PGIs cannot be translated across ancestral populations but then goes ahead and calculates his rough estimates.

7. Black men are, on average, a fraction of an inch shorter than non-Hispanic white men. However, Black women are a smidgen taller than non-Hispanic white women on average. PGIs, however, predict almost all white Europeans to be taller than almost all people of African descent. See Figure 4 of Alicia R. Martin, Christopher R. Gignoux, Raymond K. Walters, Genevieve L. Wojcik, Benjamin M. Neale, Simon Gravel, Mark J. Daly, Carlos D. Bustamante, and Eimear E. Kenny, "Human Demographic History Impacts Genetic Risk Prediction across Diverse Populations," *American Journal of Human Genetics* 100, no. 4 (April 6, 2017): 635–49.

8. The peal of *The Bell Curve* was not just heard by white supremacists, however. It was also heard by scientists such as me who wanted to know the true impact of nature and nurture. I had always dismissed *The Bell Curve* because Herrnstein and Murray didn't have genetic data. But in 2013, the Social Science Genetics Association Consortium (SSGAC), on which I served as an advisory board member, published a paper that represented the largest genome-wide association study to date. It was also the first effort to develop a PGI for education. This PGI predicted cognitive performance better than the best PGI for cognitive performance itself did. I realized that I could use the PGI for education to better test some of Herrnstein and Murray's claims. The resulting paper, written with Benjamin Domingue of the Stanford University Graduate School of Education, tested three claims made by Herrnstein and Murray:

   1. The effect of genes is increasing over time with the rise of a meritocratic society.
   2. We are becoming increasingly unequal genetically thanks to genetic elites marrying elites to a greater degree than ever before.
   3. Society is becoming less intelligent at the genetic level since those with lower ability tend to have more children than those with high cognitive ability.

   We had data from the Health and Retirement Study, which had genotyped their respondents and which included people born between 1919 and 1955—a much better time span to evaluate an argument about changes due to modernization in the twentieth century than Herrnstein and Murray's seven-year data span. Normally I might have roped more people into the research effort, but I knew that most of my colleagues, even those engaged

in sociogenomic research, would rather not have their name on a paper entitled, "The Bell Curve Revisited: Testing Controversial Hypotheses with
Molecular Genetic Data."

In short, we didn't find much evidence supporting their theses. The effect
of the PGI on educational attainment was, in fact, decreasing for more recent
birth cohorts—the exact opposite dynamic than *The Bell Curve* had argued.
It was true that the educationally endowed tended to marry the educationally endowed, but that rate wasn't increasing over time. It was stable. And it
was true that the less educationally endowed tended to have more children
than the genetically advantaged, but this, too, showed no upward trend. The
results showed that PGIs and the new science of sociogenomics weren't necessarily just going to echo the kind of claims Herrnstein and Murray were
making; sometimes they would refute those very same arguments. Maybe
sociogenomics is not just old wine in new bottles.

9. Much of the squeamishness about human social genetics, I think, does not
merely stem from the fear of eugenics, but discomfort about the specialness
and nonmateriality of the human soul being dethroned. (Circumstantial evidence to support my hypothesis comes from the fact that Confucian societies in East Asia tend to be a lot more comfortable with the exploration of
genetic bases of social life than Western populations are.)

10. The actual approach was called *instrumental variables regression*, where the gene
was used to predict alcohol consumption and, in turn, that predicted alcohol
use (rather than the observed consumption) was used to predict mortality.
See: I. Y. Millwood, P. K. Im, D. Bennett, P. Hariri, L. Yang, H. Du, C. Kart-
sonaki, K. Lin, C. Yu, Y. Chen, and D. Sun, "Alcohol intake and cause-specific
mortality: conventional and genetic evidence in a prospective cohort study of
512,000 adults in China," *The Lancet Public Health*, 8(12), 2023: e956–e967.

11. Indeed, many people associate the nature-nurture debate with an argument
of whether we have free will or not. The logic goes like this: if who we become
and everything we do is "controlled" by a script of letters in our DNA, then
we ultimately do not have agency, and we don't exercise free will. But if we
step back for a moment, we will realize that in the extreme case where we are
entirely shaped by our environment, the same argument can be made—as B.
F. Skinner made clear. The position we occupied in our mother's uterus, the
circumstances of our early childhood, the random accidents that befall us
throughout life—these outside environmental forces can shape us in a way
that eviscerates any notion of free choice, just as a DNA blueprint can. In
truth, free will is a subjective experience of consciousness, and we can leave
it to philosophers to answer whether we are just acting out a script that has
been determined since the Big Bang or whether we are each the architect of
our own destiny. From my point of view, the question of whether it's our DNA

or our environmental inputs that explain variation between us is, in essence, irrelevant to the question of whether any of us actually choose anything.

This seems like an esoteric point; however, this idea—that if genetics primarily influences who we become, we are not responsible for what actions we take—actually makes it into public discourse and even into the legal system. There are many traits that, when we hear they have a genetic basis, we end up being more empathetic toward. Take obesity. When people are told that being overweight is largely in someone's genes, they tend to be less morally judgmental about obesity. The same has been true for sexual orientation.

Back in 1970, a survey of the U.S. population revealed that 43 percent of people thought that young gay people adopted their sexual orientation due to "recruitment" by older homosexuals. In other words, Americans viewed homosexuality like the Boy Scouts or the Rotary Club—a group that people chose to join due to social influence. Perhaps not surprisingly, social acceptance of homosexuality was very low during this epoch, despite the otherwise libertine social norms of the era. But then in 1991, researchers compared identical twins and fraternal twins, as well as adopted brothers, and determined that genetic differences explained a large share of sexual orientation. The heritability of male homosexuality was estimated to be somewhere between 31 percent and 74 percent, the wide range a result of the methods used in the study. If a heritability is zero, it means that genetic differences do not explain any differences in the trait in the population; if heritability is one, then genetic differences explain all the variation in the population. So, the authors of this study, J. Michael Bailey and Richard C. Pillard, found that somewhere between a third and three-quarters of sexual orientation in men was driven by genetic differences. Later studies with better, more representative samples, put the figure in the 30–40 percent range—even if that leaves plenty of room for "recruitment" or other social forces, the notion that sexuality was even partly innate shifted attitudes toward greater tolerance.

National Institutes of Health (NIH) researcher Dean Hamer read Bailey and Pillard's study the following year and got to work trying to locate the specific genes that drove sexual orientation. After recruiting gay male patients from an NIH clinic, he constructed family trees with information about the sexual orientation of the various cousins, uncles, and so on of the focal, or proband, subject. It appeared that there was a higher degree of same-sex orientation on the maternal side, leading Hamer to study the X-chromosome. Women inherit two X chromosomes—that's what makes them female—while men inherit one X and one Y. The X comes from their mothers and the Y from their fathers. So, Hamer and his colleagues genotyped the X chromosome of his study subjects and found that the area of the X that the gay men were most likely to share clustered around a particular region: Xq28.

The mainstream media had a field day with this newly discovered "gay gene." Hamer's work combined with other studies that found differences in average brain structure between gay and straight men (specifically, hippocampal volume) to fundamentally alter the public's understanding of sexual orientation. While LGBTQ activists were initially wary of the research findings, many soon embraced this science, claiming that it proved that sexual orientation was innate, genetic, and not a matter of choice. At the time, as a young social scientist, I was quite surprised by this positive reception of biological causes by the gay community. I had been taught that biological, essentializing explanations were the first step on the road to pathologizing a group, or worse, to claiming a biological basis for a new eugenic regime. It turns out I was wrong. Since the 1990s we have seen one of the most rapid shifts in public attitudes toward nonheteronormative behavior and identity. Queer people have not only gained more public acceptance in the past three decades, but they have also won hard-fought civil rights such as to serve in the military and to get legally married. While correlation is not causation, it's pretty interesting that societal attitudes became more tolerant just as biological explanations for sexuality were waxing—especially since surveys back up the claim that once people see sexuality as biologically based, they are more accepting of nonheterosexual identities. It turns out that for many traits, when we are told that it's "in our genes," we are much more accepting and tolerant—with some notable exceptions.

12. To explain what I was intending, I asked the physician if she had ever seen the 1997 movie *Gattaca*, starring Uma Thurman and Ethan Hawke, wherein the elite of society all used artificial reproduction to create and test embryos for their various genetic risks and potentialities before deciding which to implant. "It's still you," the genetic counselor tells the parents of the protagonist. "It's just the best of you." He adds: "You could conceive a thousand times the old-fashioned way and never get a child with as good a genetic profile as we can assure you in a single round."

13. I was left to assuage my fears by reading more literature. There were many theories about why older fathers conferred more psychopathology risk to their offspring. First, there was the idea that over time, locus-by-locus mutations in the cells that produce the sperm accumulated, and research has shown that *de novo* (i.e., novel) mutational load predicts all sorts of bad outcomes. Second, there was a theory of *epigenetic mutations*—that environmental exposures lead not just to actual changes in the DNA sequence of sperm but to marks on the sperm's DNA that activate or deactivate certain genes when the egg is fertilized. Smoking and alcohol were known to cause these epigenetic marks, as were many other toxic substances. Some research implied that even psychological stress caused them. I never smoked, hardly

drank, and abstained completely from alcohol for the months leading up to my scheduled sperm extraction. The third theory, however, is what assuaged my fears. This hypothesis held that men who didn't have a child until, say, their late forties or fifties, were self-selected to be at subclinical genetic risk for psychopathology, and thus their children, by the luck of meiosis, might draw a bad hand from a risky deck and get pushed over the edge into diagnosable pathology.

In other words, it was "oddball" men that ended up not marrying and reproducing until later in life. Research supporting this hypothesis divided up the men having kids later in middle age into two groups: those for whom it was their first child and those who, as in my case, had sired children earlier in life as well. It turned out that most (but not all) the excess risk associated with advanced age lay among the first-time dads. In other words, it turned out to be mostly genetic selection and not a true treatment effect. There might even be residual genetic selection on those fathers who, like me, failed at a first marriage. But absent a great instrumental variable for having a second batch of kids with a new partner, I would never know. Subsequent research—using polygenic indices—confirmed that first-time older fathers had genetic profiles that were at higher risk for autism, ADHD, and so on.

14. A startup called Genomic Prediction led by a physicist who works in the sociogenomics space, Stephen Hsu of Michigan State University, screened five embryos for two would-be parents, the physicians Thuy and Rafal Smigrodzki. Genomic Prediction calculated polygenic indices for various traits, and the parents then chose the one they deemed to have the best genetic health profile. In late 2020, Aurea Smigrodzki became the first PGT-P baby ever born—thirty-eight years after the first IVF baby came into the world. (PGT-P stands for preimplantation genetic testing for polygenic disorders.)

"It was really a no-brainer. If you can do something good for your child, you want to do it, right? That's why people take prenatal vitamins," Rafal said. The Smigrodzki parents are, to say the least, what you would call techno-optimists. Case in point: Rafal has also signed up to have his brain cryogenically frozen upon his death in the hopes that it will one day be resurrected with his consciousness and memories intact. PGT-P was merely logical for him: "It's like the first time someone ever made a phone call—sure, it was a unique moment, but really it was just the beginning of something that now everybody does. In 10 years' time, this kind of polygenic testing will be completely non-controversial. People will be doing it as a matter of course." Aurea will be getting close to kindergarten by the time this book publishes.

I am sure Rafal is right in his claim that he and Thuy will not be alone. Though no data are available on how much business Genomic Prediction and other similar startups are doing, there is little doubt that embryos are being assayed for their risks and promise as you read this. However many are being assayed, the number is only going to increase. Meanwhile, another technology is under development that—when combined with PGI embryo selection—will be a gamechanger for reproduction: sometime in the not-too-distant future, scientists will be able to turn cheek cells into egg cells, so IVF will be painless and the number of embryos (and PGIs) to choose from will be almost limitless.

Currently, if I can have a baby naturally, it might not be worth the slight reductions in genetic risk (or increase in genetically desirable traits) that I would get from pursuing IVF. That's because during most cycles, parents get at best a dozen to twenty viable embryos. Many couples do much worse than that. And going through IVF is not only a huge painful hassle for the woman—with daily shots culminating in a surgical procedure—some scientists worry that the high doses of hormones have negative health consequences. They may even be associated with higher cancer rates.

But what if getting pregnant through IVF only involved scraping a collection stick across the inside of one's cheek or donating a vial of blood? Already, in mice, scientists can turn epithelial cells (those from inside the cheek) into pluripotent stem cells and then into ova that have resulted in viable mouse pups. *Stem cells* are cells that have not developed into their final state as neurons or white blood cells or osteocytes or, for that matter, ova. They are of huge promise for treating diseases. I, for example, am hearing impaired and check weekly as to whether there are stem-cell or gene-therapy clinical trials to restore the cells in my cochlear that are responsible for sensing sound.

There are millions of cheek cells in each swab. If even a small percentage could be turned reliably into eggs as they have been in mice, then the sky's the limit in terms of polygenic selection of embryos. Right now, the limiting factor for PGI-based embryo selection is how few eggs can be retrieved from a woman's ovaries. But if hundreds of thousands of ova could be painlessly made to match with the millions of sperm in most ejaculations, that's an entirely new ballgame. When ova are retro-engineered from skin cells, then all bets are off, since prospective parents will be able to choose not among a dozen or so embryos, but among hundreds of thousands or even millions.

Choosing the best of twelve might provide some marginal benefit in terms of lower disease risk or greater educational promise. But the laws of statistics say that the embryos are not likely to be very far apart in their scores. By contrast, when there are thousands of random draws, the bell curve starts to fill in and rare outcomes in the left and right tails can be sampled. When parents can create thousands of embryos, they will have a much wider, and more statistically meaningful, range of potential children to choose from.

First, this almost unlimited choice means that if a given couple is obsessed with maximizing a single trait or minimizing genetic risk for a single disease, they will much more effectively be able to do that. Depending on their own PGIs, they may not be able to get down to the lowest fifth percentile of risk, but they will be able to get to, say, twentieth percentile risk if they are starting at an average fiftieth percentile of risk for themselves. If they are both at the seventy-fifth percentile for the education or cognitive ability PGI, they may be able to choose an embryo at the ninety-fifth percentile.

The geneticist Peter Visscher once joked that if we switched every known allele to the version that made someone taller, that person would be predicted to be over twenty feet tall. What he meant was that if you look at the effect of each allele in a GWAS, gave someone two of the "taller" alleles, and then added them all up to get a predicted height, you would have constructed a genetic giant. Now, that won't happen in embryo selection for a few reasons. First, nobody will pull a perfect straight flush across thousands of draws—especially since, in the case of embryo selection, we are limited to choose genetic cards from those the parents are carrying in their respective decks. Second, nobody would actively choose a baby predicted to be twenty feet tall. Third, PGIs usually predict best in the mid-range of the distribution since that's where most of the people are who they were trained to predict in the GWAS that went into their construction. As we get far out, the assumptions of a step-by-step linear increasingly break down; we cannot assume when someone is already predicted to be seven feet tall that adding another predicted 0.1 inches will have the same effect as if we added that 0.1 inch to someone who was five-foot-ten. In some cases, there are physiological constraints (like bone density) in building a twenty-foot human. (By contrast, for other cases, risk actually increases more at the tail end. This is the scenario for the right tail of the polygenic index for heart disease. The risks for heart attacks start to multiply as you move from the ninety-fifth percentile to the ninety-ninth percentile in the PGI rankings.)

Besides being able to screen for a single trait more effectively, when parents have hundreds of thousands of embryos to choose from, they will face fewer tradeoffs. What I mean by this is that if they are set on getting an embryo in the top 10 percent for education and a child in the bottom 10 percent for heart disease, they will not have as much trouble finding one that meets both those criteria rather than having to face a Hobson's choice.

Inevitably, such precision selection will still present risks. Given widespread *pleiotropy*, the fact that genes have multiple effects, maximizing on one dimension may introduce other risks that were not anticipated. For instance, maximizing on cognition may result in a highly myopic baby given a genetic correlation between the two. Extreme myopia, in turn, is a risk factor for glaucoma, retinal detachment, and macular degeneration—three different precursors to blindness. And that's just one pleiotropic effect we know about. There may be many more lurking in the genetic shadows.

And that's not even considering how sperm and ova banks will change how they do business thanks to incorporating PGIs into their selection regimes. Nor how dating apps might hoover up genetic information. Perhaps the lowest hanging fruit involves sperm banks and the market for egg donation. Right now, most sperm banks test for only a few inherited diseases, and the same is true for the screening of egg donors. If I were entrepreneurial, the first business I would start would be a sperm bank that tests the donors, calculates PGIs for a whole range of outcomes, and provides that information to prospective clients. This seems like it should happen any day now (if it's not already happening). But the consequence is that sperm will become a winner-take-all market in which everyone wants the sperm of just a few men who score highly on desirable PGIs.

Humans don't currently arrange their own mating like we do for thoroughbred horses, with all the focus on bloodlines, at least not explicitly. But once we get to know our PGIs for a range of outcomes—and more importantly, those of potential mates—it may be hard to keep selective breeding sensibilities unspoken. After all, we are already genetically sorting without any explicit information about our DNA. When we have all measured our own and our partners' PGIs, mating and dating will look a lot more like thoroughbred horse breeding.

Before we even talk about genetically selected babies, there are potential ethical entanglements waiting that could spring up before a couple even have a first date. Today, several new startups (DNA Romance, for instance) are using genetic information to offer potential romantic matches. DNA Romance doesn't just test the DNA of existing couples, it offers an online dating marketplace where "compatibility" measures are computed for the profiles you see online. These compatibility measures are based on some old-fashioned factors like personality tests and photographs but add the new wrinkle of DNA into their algorithm. DNA is assessed for the major histocompatibility complex (MHC), an area of chromosome 6 involved in natural immunity to disease for which it is healthier to maintain diversity to protect against novel diseases. DNA Romance also assays DNA to see if someone is a distant genetic relative (steering you away from them), and the company calculates polygenic scores for a number of phenotypes, adding those into the score with a dash of salt.

In the U.S., accidental marriages between second cousins are rare, given how large and diverse our dating markets tend to be—with the notable exceptions of some small communities like the Amish or ultra-Orthodox Jews. But in Iceland, an island population descended from a small group of initial settlers, DNA relatedness is rife. There, there's a DNA-based app called, excuse my Icelandic, "Wait, Wait, Don't Fuck Her, She's Your Cousin." Meanwhile, some Orthodox Jewish yentas are using genetic information to avoid recessive gene combinations.

The current panoply of genetic dating apps are just curiosities bordering on scams. (Perhaps DNA Romance serves an important function of screening for genetics nerds like me, creating a small community of likeminded geeks.) But I envision a more seamless integration of genetic information, PGIs in particular, into the current ecology of dating apps—Tinder, Bumble, Grindr, Hinge, and so on. Maybe it starts with an entrepreneur and some angel investors, but it could also take off if/when users, themselves, voluntarily post their own PGI information on their existing profiles.

Of course, love is not logical at the individual level, but collectively, we act as rational maximizers when it comes to dating and mating. Analysis by sociologist Elizabeth Bruch found a distinct pattern in online dating: people with similar ratings ended up together. But how this happened was the interesting part of the story. It's not like the fiftieth-percentile most-desired man preferred the fiftieth-percentile most-desired woman on the site (and vice versa). They both preferred the ninety-ninth-percentile-rated person on the site. They may have tried messaging or swiping right on that profile only to find themselves ignored. Working their way down the list, so to speak, they settled for each other. There's no reason to expect genetics to work any differently. If you found yourself confronted with two equally attractive suitors, one who was at the ninety-fourth percentile for risk of schizophrenia and the other who was at the sixteenth percentile, which would you message back?

Once the initial ickiness factor falls away (which it will), companies like 23andMe or Ancestry might offer a certification of those scores—the way X (formerly Twitter) used to have blue checkmarks for verified users. The discomfort of oversharing will, I believe, fade because people will see the stakes as high enough—that is, the potential risks and traits of one's offspring. I can see the ad copy already: "Phenotype is for hookups, genotype is forever." In a decade, mating and reproduction may be unrecognizable to us today.

15. Throughout this discussion, you might have been wondering: What about CRISPR gene editing? Won't the ability to edit our own genomes make all this hullabaloo about polygenic prediction moot? Indeed, much more ink

has been spilled about the potential for CRISPR gene editing technology to radically transform our lives on multiple fronts than there has been about genetic prediction and social genomics. In some ways this is understandable. This gene-editing tool is already improving crop yields and reducing reliance on pesticides. It has also provided hope for those suffering from crippling diseases. For instance, people with sickle cell anemia can have their stem cells edited to remove the mutated HBB gene and then have their bone marrow irradiated and repopulated with the repaired cells, essentially curing their disease. This technique can be used for several other conditions as well, including AIDS.

It is relatively uncontroversial to edit somatic cells—cells that are within the body but whose DNA is not passed on to future generations. But so-called *germline gene editing*—when the cells that produce ova or sperm, or more likely, embryos themselves, are edited—is another story. The altered version of the gene would appear not only in every cell in the patient to be born but also in their offspring. It is germline applications of CRISPR that inspire the most sci-fi, utopian (or dystopian) visions: visions of a world where sickle cell anemia is gone once and for all, along with Huntington's disease, Tay-Sachs, and most other Mendelian disorders; a world in which schizophrenia and hypertension become rare, and perhaps we become a bit smarter, taller, and faster. (Indeed, a National Academy of Sciences, Engineering, and Medicine panel has called for a moratorium on germline gene editing using CRISPR or related technologies.) The reality of reproductive genetics, however, is that genetic editing of embryos, ova, and sperm (or their progenitor cells) is not practical for most conditions of interest and, actually, not even necessary.

For most single gene diseases, it's simpler and less risky to just select embryos that don't suffer from those conditions. Meanwhile, most other phenotypes—from cardiovascular disease to cognitive ability—are influenced by so many tiny genetic differences spread across all twenty-three pairs of chromosomes that CRISPR isn't feasible. The bottom line is that—notwithstanding the work of the Chinese scientist He Jiankui, who used CRISPR to make two human embryos immune to HIV—we are going to have a lot more Aureas before we have more CRISPR babies. (Jiankui had thought he would be viewed as a scientific hero, but instead his actions were seen as medically reckless, especially since established ways to prevent babies from contracting HIV from their mothers both in utero and during birth already existed. He was convicted of breaking domestic law and sentenced to three years in prison.) Genetically selected human organisms will be the norm long before GMO humans are.

Chapter 2

1. Other Greek thinkers, however, offered a more nuanced account. Aristotle, Plato's student, saw humans as having predispositions and potential at birth but believed that the civilizing process was necessary to bring those to the fore—giving a more important role to environment than Plato had afforded it. Likewise, Hippocrates, considered the founder of Western medicine, saw an interplay of nature and nurture in human health and development. He believed in a *physis*—a natural constitution—that everyone possessed. Yet he also saw a role for climate, soil, food, and social conditions in shaping the health and wellbeing of humans. In this vein, he saw the physical features of what we would call race in today's parlance as something that resulted from specific combinations of soil, water, and air. In this way, Hippocrates may have been the first to lay out a theory of gene-environment mutual dependence.

    These questions of the relative impact of innate characteristics and environmental influences occupied Renaissance and Enlightenment thinkers in the West as well. For instance, in the early seventeenth century, French philosopher René Descartes emphasized the predetermined nature of human beings. In 1639, he argued that certain fundamental aspects of the human mind are ingrained from birth. Among these innate features were our capacity for reason and an inherent sense of the existence of God. These inherent notions form the foundation for acquired knowledge that help humans comprehend the world. To take a metaphor from modern linguistics: all children are born with the hardware that allows them to learn language. It's just the software (i.e., their particular dialect) that is determined by the environment. Descartes's schema can be seen as akin to that of Hippocrates or Aristotle: nature lays the groundwork for nurture.

2. He was inspired by a journey he took to North Africa almost two decades earlier.

3. More recent research suggests that non-African *Homo sapiens* actually did interbreed with other humanoid species—Neanderthals and Denisovans.

4. F. Galton, *Inquiries into Human Faculty and Its Development* (London, 1883), 307.

5. Margaret Sanger, *The Pivot of Civilization in Historical Perspective* (New York: Brenato's Publishers, 1922), 80, 82, 176, as quoted in Abigail Shivers, "Margaret Sanger: Ambitious Feminist and Racist Eugenicist." *Woman Is a Rational Animal* (blog), University of Chicago, September 21, 2022.

6. Sanger, *The Pivot of Civilization*, 101–102.

7. Minnesota Governor's Council on Developmental Disabilities, "Rise of Eugenics," from *Parallels in Time: A History of Developmental Disabilities, Part I:*

The Ancient Era to the 1950s, the Rise of the Institutions 1800–1950, Online Historical Archives, Mn.gov, n.d.

8. Despite counting the great Man o'War as his great-grand sire, the rest of Kelso's pedigree was relatively undistinguished.

9. A. K. Thiruvenkadan, N. Kandasamy, and S. Panneerselvam, "Inheritance of Racing Performance of Thoroughbred Horses," *Livestock Science* 121, no. 2–3 (April 2009): 308–26.

10. K. S. Kendler, L. M. Karkowski, and C. A. Prescott, "Fears and Phobias: Reliability and Heritability," *Psychological Medicine* 29, no. 3 (May 1999): 539–53.

11. Seema Mehta, "Trump's Touting of 'Racehorse Theory' Tied to Eugenics and Nazis Alarms Jewish Leaders," *Los Angeles Times*, October 6, 2020.

12. Brandon Tensley, "The Dark Subtext of Trump's 'Good Genes' Compliment," CNN, September 22, 2020.

Chapter 3

1. C. Wright Mills, *The Sociological Imagination* (London: Oxford University Press, 1959).

2. I was essentially comparing two children from families that were pretty much identical on their incomes, their parents' education levels, their parents' jobs, the number of children in the household, the age at which the parents had them, and so on, but who were different in their families' net worth. If they were identical on all those other dimensions, from where did the wealth differences between these two families arise?

It could be that luck, otherwise unrelated to anything fundamental about those two families, caused the wealth differences. One family could have won the lottery, or had a long-lost rich uncle who bequeathed the parents his estate, or any other fortuitousness. Or it could be that I was not measuring some secret ingredient. Perhaps one set of parents shared a culture of thrift and savings while the other family, even though they achieved the same levels of education, same jobs, and same income, were more hedonistic and liked to spend their earnings on vacations and fancy wine. Or it could be that in one family, the parents were much more mathematically oriented and understood the stock market and finance in general much better, and thus they managed to accumulate a much greater return on their investments.

Parental hedonism or mathematical ability or investment savvy would all be likely to have a direct effect on children's outcomes independent of their effects on the net worth of the household. It is not a stretch of the imagination to think that parents who model hedonistic behavior for their children will socialize those children differently than those who model thrift and planning—regardless of their wealth levels. This meant that the answers I was getting from my analysis, which ignored these factors, were likely wrong (*biased*, in the technical term of the field).

By way of a solution to this problem, I could try to beg the Panel Study of Income Dynamics (the dataset I had used for my research) to ask the survey subjects about these more intangible, psychosocial aspects of a family and add them to my statistical models to factor them out of the analysis. However, even if I could go back in time to ask them in 1984, such concepts like "hedonism" or even "investment savvy" are notoriously hard to pin down and measure with any semblance of accuracy. Moreover, this would be a Sisyphean task because each time I had ruled out one potential factor that could explain why some families had more wealth than others and that also might impact the children's outcomes directly, someone could come up with another. Multiple regression analysis, then, was like a game of whack-a-mole, where the list of potential variables to factor out was seemingly boundless.

3. Siblings can range in actual relatedness anywhere from about 35 percent to 65 percent by chance. Perhaps some of the fraternal twins who were on the upper end of that distribution—being close to two-thirds identical rather than half—seemed too alike to *not* be identical. Likewise, identical twins can look different for a whole variety of reasons. One twin might have gotten more nutrients in utero just by virtue of having gotten a front row seat to the placenta, thereby hogging maternal nutrients. This twin, in turn, would be born heavier and would probably retain a height advantage their whole life. Other random differences, from the epigenome outward to skin tone, can fool people about their children or themselves. A birth mark here, a mole there, lighter shading, and so on can all be random differences between twins. As it turned out, it was much more likely for identical twins to be misclassified as fraternal than for dizygotic twins to be taken as monozygotic.

4. Such a finding would not completely disqualify the EEA since it could be that they were less similar simply because of whatever differences they experienced that led them to be misclassified in the first place.

5. Dalton Conley, Emily Rauscher, Christopher Dawes, Patrik K. E. Magnusson, and Mark L. Siegal, "Heritability and the Equal Environments Assumption: Evidence from Multiple Samples of Misclassified Twins," *Behavior Genetics* 43, no. 5 (September 2013): 415–26.

6. They examined 2,748 publications including 14,558,903 partly dependent twin pairs. (There have not literally been 14,558,903 pairs of twins studied. Rather, *partly dependent twin pairs* means that many of the studies used the same samples of twins; the total number of times twins have been studied is over 14 million.) Tinca J. C. Polderman, Beben Benyamin, Christiaan A. de Leeuw, Patrick F. Sullivan, Arjen van Bochoven, Peter M. Visscher, and Danielle Posthuma, "Meta-Analysis of the Heritability of Human Traits

Based on Fifty Years of Twin Studies," *Nature Genetics* 47, no. 7 (May 2015): 702–709.

7. See, for example, David Cesarini, Erik Lindqvist, Robert Östling, and Christofer Schroeder, "Does Wealth Inhibit Criminal Behavior? Evidence from Swedish Lottery Winners and Their Children," NBER Working Paper, 31962.

8. See, for example, Daniel J. Benjamin, David Cesarini, Christoper F. Chabris, Edward L. Glaeser, David I. Laibson, Vilmundur Guðnason, Tamara B. Harris, et al., "The Promises and Pitfalls of Genoeconomics*," *Annual Review of Economics* 4 (July 1, 2012):627–62. https://doi.org/10.1146/annurev-economics-080511-110939.

9. Gregory Clark, "The Inheritance of Social Status: England, 1600 to 2022," *Proceedings of the National Academy of Sciences* 120, no. 27 (June 2023): p.e2300926120.

10. Kenneth Wong, Crystal Thomas, and Megan Boben, "Providence Talks: A Citywide Partnership to Address Early Childhood Language Development," *Studies in Educational Evaluation* 64 (2020): 100818.

11. Shawn E. Christ, "Asbjørn Følling and the Discovery of Phenylketonuria," *Journal of the History of the Neurosciences* 12, no. 1 (2003): 48.

12. Christ, "Følling and the Discovery of Phenylketonuria," 48.

13. Office of the Press Secretary, "June 2000 White House Event," White House press release, June 26, 2000.

14. The mechanism seemed to be the fact that people with the A—which came to be known as the Taq1A polymorphism—had 40 percent fewer dopamine receptors present in the corpus striatum, the keystone of the brain's reward system. Fewer receptors, in turn, were theorized to translate into a higher threshold for pleasure—hence the need for these individuals to give themselves more of a rush.

15. These sites of common variation are called *single nucleotide polymorphisms* or *SNPs*, for short. Other forms of variation are *indels* (insertions or deletions), *copy number variants* (*CNVs*), and rare or *de novo* variants. SNPs are what are usually studied.

16. Louis I. Woolf and John Adams, "The Early History of PKU," *International Journal of Neonatal Screening* 6, no. 3 (July 29, 2020): 59.

17. Despite the smashing success of this environmental intervention, it remains the exception rather than the rule. Hereditary Mendelian diseases have occupied one lane of medicine and environmentally caused diseases (such as infections) another. These parallel paths have typically not converged. While the results from Woolf's dietary intervention were accepted by the medical community as real and valuable, they did not lead to a sea change in thinking about genes and the environment. Today if you search for treat-

ments for genetic disorders, you get a lot of information about gene thera-
pies, bone marrow transplants, and medicines to manage symptoms. And it
is true that nobody has found a fix for Huntington's disease or for Tay Sachs
or for most other single-gene hereditary diseases, even if the hope is that
gene editing and replacement therapy will soon provide a cure.

But, as the polygenic revolution has demonstrated in the past fifteen
years, those single-gene diseases are the exception and not the rule.
When it comes to polygenic traits and diseases—running the gambit from
Alzheimer's to hypertension to depression—the interplay of genes with the
environment is beginning to become accepted. Yes, we have long known
that something like high blood pressure is a result of both genetic factors—
how efficiently your body pumps sodium out, how naturally elastic your
blood vessels are, and so on—as well as environmental ones—how much
sodium you consume, your stress level, your exercise habits, and so on.
But it's only recently that we've seen that the lanes have merged—that for
many hereditary diseases the interplay of our genes and our environments
is what matters. There are probably many reasons why we have collectively
resisted the braiding of nature and nurture in this way. We like straight-
up explanations. It's not as satisfying to say—about the very genes in our
cells—well, it depends.

18. While PGIs don't come close to reaching the same estimates of heritability
as the old twin studies, even explaining a fraction of those huge tranches is
incredibly useful for studying the social world. Moreover, while the PGI
does not pick up the entire effect of genes, there are ways of extrapolat-
ing what would happen if it did—specifically a technique called *SIMEX*, or
simulation-extrapolation (of measurement error).

19. By way of example, a very exciting current area of research in human health
and behavior is called *epigenetics*. Epigenetic studies in humans often examine DNA *methylation*—or the addition of chemical tags (specifically CH_3) to
particular locations across the genome—to study gene regulation. Regions
of your genome that are highly methylated tend to be shut down (i.e., the
proteins they encode are not produced), and those that are less methylated
tend to be *expressed*—meaning they're on and making proteins such as hormone receptors, collagen for your skin, and so on. What's exciting about this
domain of research to many behavioral scientists is the fact that aspects of
our environment seem to influence methylation. DNA methylation gives us
a window into how the social and physical environment gets under our skin
and into our cells, and it actually affects the expression of our genes. If a
stressful home environment or a neighborhood with toxic air quality stunts
physical growth, we can now see which genes are being repressed and which
are being expressed in the pathway between, say, parental yelling/inhaled

particulate matter and short stature. We've already known for years that gene expression is affected by environmental influences, but DNA methylation is a particularly easy way to measure these effects.

These data are very exciting to many social and behavioral researchers since they demonstrate that the environment really does matter—so much so that we can read its fingerprint in someone's DNA. It follows a causal arrow from the environment around us into our genome. What better proof that nurture is a powerful force than the fact that it annotates the very script of who we are? By contrast, social scientists tend to be a bit queasy about sociogenomics out of the (mistaken) concern that if outcomes are partly written in our fixed biological blueprint, there is nothing we can do to improve them. Or worse, that they are wholly biologically programmed in a deterministic manner, or worst of all, that acknowledging the influence of genes on social life is the first step down the road to eugenics.

In turn, how someone's genome is methylated can tell us how rapidly or slowly they may be aging as compared to their chronological age, how they may respond to pathogens, or even how emotionally reactive they may be. DNA methylation is just one example of an explosion of biomarker data. Other biological measures that are currently being studied to detect bodily embodiment of the environment include telomere length (how short or long the ends of our chromosomes are, another indicator of cellular aging), C-reactive protein levels (a marker of inflammation that supposedly captures stress), and even hematopoietic aneuploidy (as an indicator of cumulative exposure to toxins, among other things). All these and many more biological indicators are no doubt exciting sources of information for researchers, but they are all qualitatively different from DNA in that they are both *cause* and *effect* of the social dynamics we want to study. If smokers display increased methylation around a certain gene, did that cause them to smoke? Or is that signature a result of the chemicals they are inhaling through their lungs? Keep in mind that such a riddle may be solvable through other means, but nothing about epigenetic data per se helps solve it. We are back to the old search for experiments to establish cause and effect. Not so for DNA. If we find a certain genetic—as opposed to epigenetic—profile common to most smokers as compared to non-smoking individuals of the same ancestry (ideally siblings), we can be certain the genotype is causing the smoking (though we may not know for certain how). It can't be the reverse.

20. Social and behavioral scientists can use PGIs in their research in at least two straightforward ways: by reducing bias and by increasing statistical power. First, since most outcomes are heritable, and since genotypes are not randomly distributed across environments, most observational estimates of

environmental effects are biased. This is the classic problem of the lurking unobserved variable—genetics—that plagued Susan Mayer (and later me) with respect to poverty research. By having a measure that quantifies genetic risk for, say, education, we can see how the estimates of parental poverty (or wealth) on offspring education change (or not) when we add the education PGI. Second, even if we have a randomized controlled trial (or natural experiment) where we know genes are randomly distributed across the treatment and control groups thanks to the experimental design, by including the relevant PGIs in our statistical comparison between the two groups, we increase what statisticians call *power*. That is, the PGI takes out some of the *noise*—the variation in people's outcomes that has nothing to do with the experiment—and thus makes the signal stronger and easier to detect. This, in turn, allows us to discover more subtle treatment effects with smaller samples.

Chapter 4

1. President Franklin Delano Roosevelt's health had always been a challenge, but by his final year in office, it had become extremely poor. Among his problems were heart ailments, high blood pressure, and bronchitis. On April 12, 1945, FDR died of a cerebral hemorrhage. Even though he had been ailing, his death was still a shocking blow to the U.S. and the world. Partially in response to FDR's death, Congress passed legislation to study the roots of heart disease.

2. So little was known, in fact, that the initial budget allocation of $94,350 included a line item for ashtrays for the smoking needs of the staff.

3. They ended up collecting data on approximately six thousand of the roughly ten thousand adult residents. Syed S. Mahmood, Daniel Levy, Ramachandran S. Vasan, and Thomas J. Wang, "The Framingham Heart Study and the Epidemiology of Cardiovascular Disease: A Historical Perspective. *Lancet* 383, no. 9921 (March 15, 2014): 999–1008.

4. Excepting Sir James Mackenzie's aborted attempt to longitudinally follow residents of St. Andrews, Scotland.

5. T. R. Dawber, F. E. Moore, and G. V. Mann, "Coronary Heart Disease in the Framingham Study," *American Journal of Public Health and the Nation's Health* 47, no. 4, part 2 (April 1957): 4–24.

6. W. B. Kannel, T. R. Dawber, M. E. Cohen, and P. M. McNamara, "Vascular Disease of the Brain—Epidemiologic Aspects: The Framingham Study, *American Journal of Public Health and the Nation's Health* 55, no 9 (September 1965): 1355–66.

7. Nicholas A. Christakis, and James H. Fowler, "The Spread of Obesity in a

Large Social Network over 32 Years," *New England Journal of Medicine* 357, no. 4 (July 26, 2007): 370–9.

8. In a second article on the topic, they showed that simply considering the common environments experienced by friends greatly reduced the effect of being friends. What's more, changing the model to measure only the change in BMI from the time two individuals became friends further eroded the contagion effect. If friends were really affecting each other's weight, we would expect to see a BMI change once they befriended each other. But there wasn't one.

9. Benjamin W. Domingue, David H. Rehkopf, Dalton Conley, and Jason D. Boardman, "Geographic Clustering of Polygenic Scores at Different Stages of the Life Course," *Russell Sage Foundation Journal of the Social Sciences* 4, no. 4 (2018): 137–49.

10. Nicholas A. Christakis and James H. Fowler, "Friendship and Natural Selection," *Proceedings of the National Academy of Sciences* 111, no. 3 (2014): 10796–801.

11. Christakis and Fowler discovered some mysterious processes that are operating underneath our noses—literally—that drive this process of genetic sorting. Namely, they identified specific regions of the genome that display particularly high levels of similarity between friends. Interestingly, these regions are associated with olfaction. They speculated that perhaps smell, and pheromones in particular, were important, if unconscious, dimensions on which we decide with whom we want to spend time and who we want to avoid. Furthermore, the genes on which we tend to match with our friends tend to be under higher-than-average positive natural selection. That is, evolution favors our genetic sorting, on some dimensions, at least. That's a hard force to fight against. Meanwhile, genes associated with immune function tended to be heterophilic—meaning opposites attract. Such immunological heterophilia among mates has long been shown and posited to serve to maintain immunological diversity in the population.

12. Overall, second cousins (and U.S. adolescent friends) tend to be 3.1 percent genetically similar. (Random strangers would be 0 percent similar on average.)

13. In a related study, Barnes and colleagues examined whether the measurable characteristics of our friends were influenced by our own genetics—that is, how heritable the GPAs of our friend group is. (J. C. Barnes, Kevin M. Beaver, Jacob T. N. Young, and Michael TenEyck, "A Behavior Genetic Analysis of the Tendency for Youth to Associate According to GPA," *Social Networks* 38 [2014]: 41–49.). Harden and colleagues found this dynamic to hold for alcohol and tobacco use as well. (K. Paige Harden, Jennifer E. Hill, Eric Turkheimer, and Robert E. Emery, "Gene-Environment Correlation and Interaction in Peer Effects on Adolescent Alcohol and Tobacco Use," *Behavior Genetics* 38, no. 4 [July 2008]: 339–47.) Finally, Guang Guo found that

our genes influenced our friend group characteristics for GPA, aggressive behavior, and cognitive ability. This does not mean that my genes actually cause my friend's GPAs or drinking behavior—though that might be possible through peer effects. Most of the effect is that attracting and choosing friends, like almost every other behavior, is moderately heritable. What's more, Fowler, Dawes, and Christakis found that the specific network position an individual occupies is also related to that person's genes. (James H. Fowler, Christopher T. Dawes, and Nicholas A. Christakis, "Model of Genetic Variation in Human Social Networks," *Proceedings of the National Academy of Sciences* 106, no. 6 [February 10, 2009]: 1720–4.) Whether a given person was on the periphery of the social graph, as it's called, or at the center—where many paths run through them—was influenced by their genes. Put simply, popularity rests partly in your DNA.

14. Jonny Wilkes, "What Was the Habsburg Jaw?" HistoryExtra, the official website for the *BBC History* magazine, July 12, 2022.

15. Benjamin W. Domingue, Jason Fletcher, Dalton Conley, and Jason D. Boardman, "Genetic and Educational Assortative Mating among US Adults," *Proceedings of the National Academy of Sciences* 111, no. 22 (June 2014): 7996–8000.

16. Two important caveats are worth noting before we despair on the coming epidemic of recessive diseases. First, when we match on PGIs, two spouses could have attained those PGIs differently, making their actual similarity on a gene-by-gene basis lower than it would appear from the PGI correlations. Imagine a scenario in which you score high on the education PGIs calculated from the odd number chromosomes and your partner scores high on the even number ones, for example. Second, most genetic diseases are caused by relatively rare variants, not the type that are picked up by PGIs, which currently capture only common variation.

17. Spouses' similarity on a genetic level is actually higher than their similarity in their actual stature. This suggests that height "genes" are also correlated with other observable factors that people value or disvalue in a potential romantic partner. Such factors could be related to physical looks, but they may also lie in the domain of athleticism, general health, or some yet undiscovered factor (smell maybe?).

18. On the genetics of education, we found that U.S. spouses were more like first cousins. (Later work in the UK also found a similar dynamic with respect to education.) Interestingly, as women gain access to higher education, assortative mating on actual education level has been rising, but spousal similarity on the genetics of education has not been rising in tandem. It seems as though men and women were already sorting on women's genetic potential for education even when the opportunities were not present for that potential to be realized.

19. Abdel Abdellaoui, David Hugh-Jones, Loic Yengo, Kathryn E. Kemper, Michel G. Nivard, Laura Veul, Yan Holtz, et al., "Genetic Correlates of Social Stratification in Great Britain," *Nature Human Behaviour* 3, no.12 (October 21, 2019): 1332–42.

20. Zoya Gubernskaya and Dalton Conley, "Does Polygenic Risk Explain 'Immigrant Health Paradox'? Evidence for Non-Hispanic White Older Adults from the Health and Retirement Study" (working paper, SUNY Albany, 2023).

21. Daniel W. Belsky, Terrie E. Moffitt, David L. Corcoran, Benjamin Domingue, HonaLee Harrington, Sean Hogan, Renate Houts et al., "The Genetics of Success: How Single-Nucleotide Polymorphisms Associated with Educational Attainment Relate to Life-Course Development," *Psychological Science* 27, no. 7 (July 2016): 957–72.

22. Hamed Nilforoshan, Wenli Looi, Emma Pierson, Blanca Villanueva, Nic Fishman, Yiling Chen, John Sholar, Beth Redbird, David Grusky, and Jure Leskovec, "Human Mobility Networks Reveal Increased Segregation in Large Cities," *Nature* 624, no. 7992 (2023): 586–92.

23. One rejoinder might be that it was a different PGI that predicted BMI or smoking in the old days, and if we had more data on superannuated people, we could develop a better PGI for earlier generations. But two factors mitigate against this possibility. First, the birth cohorts that went into the prediction of BMI and smoking tend to be the older ones, so the declining predictive power for newer birth cohorts isn't likely due to ascertainment bias (see, e.g., Dalton Conley, Thomas M. Laidley, Jason D. Boardman, and Benjamin W. Domingue, "Changing Polygenic Penetrance on Phenotypes in the 20th Century among Adults in the US Population," *Scientific Reports* 6, no. 1[July 26, 2016]: 30348). Second, the overall heritability (i.e., genetic influence) seems to be increasing as well (see, e.g., Jason D. Boardman, Casey L. Blalock, and Fred C. Pampel, "Trends in the Genetic Influences on Smoking," *Journal of Health and Social Behavior* 51, no. 1 [March 2010]: 108–23.)

24. Of course, income (or education or most traits we care about from an inequality perspective) are not 100 percent heritable. And people are not perfectly assorting. Notwithstanding the seemingly unlimited pool of potential mates, thanks to online dating, in reality, our pool of matches is largely constrained by geography and other factors—including who swipes right on us. Moreover, most of us are not monomaniacally focused on a single trait like education. We are looking for someone who complements us on multiple fronts. We make tradeoffs, so even though you may tell your spouse that you could not imagine anyone more perfect than them, the truth is that we could all improve on a number of fronts. Finally, in modern society, mating is not a rational process. Yes, a pattern of sorting does go on underneath the surface

in broad terms, but people are reacting emotionally to who they encounter in their dating lives. And those emotions may lead us to people who are different than us, genetically speaking. But to the extent that people do end up with spouses with more similar genetics on a trait—not just income and education, but height, depression, and autism, to name a few—then genetic inequality is higher for the subsequent generation and mobility is lower than it would be otherwise.

25. Hans Fredrik Sunde, Nikolai Haahjem Eftedal, Rosa Cheesman, Elizabeth C. Corfield, Thomas H. Kleppesto, Anne Caroline Seierstad, Eivind Ystrom, Espen Moen Eilertsen, and Fartein Ask Torvik, "Genetic Similarity between Relatives Provides Evidence on the Presence and History of Assortative Mating," *Nature Communications* 15, no. 2641 (March 26, 2024). Technically, one doesn't need all the generations in the data to measured assortment. In each person's genome there is a record of prior generations' assortment. To wit, the simple correlation across independently sorting chromosomes in the PGI for a given trait tells us how assorted the parental generation was. Likewise, the correspondence across chromosomes with respect to higher-order statistical moments—like the variance and skewness—tell us about prior generations as well. (But there's no substitute for actually having data on multiple generations in terms of statistical power.) Using the approach based on unrelated individuals, there is evidence that Norway is not alone but that the UK genetic assortment on height and education has been increasing over the last century as well.

26. Abdellaoui et al., "Genetic Correlates of Social Stratification in Great Britain."

27. Moreover, with the current PGI technology that works only for those of European descent, racial inequality will be exacerbated as whites increasingly have access to the ability to "improve" the genomes of their children while non-whites are left to reproduce the old-fashioned way—that is, by chance.

Chapter 5

1. Bruce Sacerdote, "Peer Effects with Random Assignment: Results for Dartmouth Roommates," *Quarterly Journal of Economics* 116, no. 2 (May 2001): 681–704.

2. Michael Kremer and Dan Levy, "Peer Effects and Alcohol Use among College Students," *Journal of Economic Perspectives* 22, no. 3 (Summer 2008): 189–206; Daniel Eisenberg, Ezra Golberstein, and Janis L. Whitlock, "Peer Effects on Risky Behaviors: New Evidence from College Roommate Assignments," *Journal of Health Economics* 33 (January 2014): 126–38.

3. Sacerdote, "Peer Effects with Random Assignment," 681–704.

4. Back in the candidate gene era, University of North Carolina professor Guang Guo and his colleagues had the foresight to collect saliva from his sample of roommates. He then analyzed whether single, candidate genes affected how roommates reacted to each other. He found that for female students with a gene that is associated with being overweight (*FTO*), having a frequently exercising roommate led to lower weights by the end of freshman year than among women who had the weight-enhancing version of the gene who got a less active roommate. For women with the more protective version of the gene, it didn't matter what roommate they got: a classic gene-environment interaction effect. (Yi Li and Guang Guo, "Peer Influence on Obesity: Evidence from a Natural Experiment of a Gene-Environment Interaction," *Social Science Research* 93 [January 2021]: 102483.) They found a similar dynamic for drinking: having a heavily drinking roommate influenced those who were at a middle level in terms of propensity to consume alcohol (as measured by a primitive PGI constructed from candidate genes), but those who were at the high end or low end were unaffected by whom they were assigned. (Guang Guo, Yi Li, Hongyu Wang, Tianji Cai, and Greg J. Duncan, "Peer Influence, Genetic Propensity, and Binge Drinking: A Natural Experiment and a Replication," *American Journal of Sociology* 121, no. 3 [November 2015]: 914–54.) This makes sense. Those who are genetically hard wired to drink will probably do so anyway—no matter who they get as a roommate. And those who are genetically averse to alcohol will also probably not suddenly drink more because their roommate does. It's those in the middle, the alcohol swing voters, if you will, who are persuadable by social influence.

5. Olga Yakusheva, Kandice Kapinos, and Marianne Weiss, "Peer Effects and the Freshman 15: Evidence from a Natural Experiment, *Economics and Human Biology* 9, no. 2 (March 2011): 119–32.

6. Olga Yakusheva, Kandice A. Kapinos, and Daniel Eisenberg, "Estimating Heterogeneous and Hierarchical Peer Effects on Body Weight Using Roommate Assignments as a Natural Experiment," *Journal of Human Resources* 49, no. 1 (2014): 234–61.

7. Seth Stephens-Davidowitz, *Don't Trust Your Gut: Using Data to Get What You Really Want in Life* (London: Bloomsbury Publishing, 2022).

8. Marco Ciabattini, Emanuele Rizzello, Francesca Lucaroni, Leonardo Palombi, and Paolo Boffetta, "Systematic Review and Meta-Analysis of Recent High-Quality Studies on Exposure to Particulate Matter and Risk of Lung Cancer," *Environmental Research* 196 (May 2021): 110440.

9. There are no same-sex marriages in the Health and Retirement Study, the dataset which we examined.

10. *Sunderland Daily Echo and Shipping Gazette*, October 22, 1903.

11. With the usual caveats about measurement error that could attenuate our

estimates. Even if we multiplied the upper bound of the estimate by a factor of three to account for the ratio of the PGI's explained variance to the full SNP heritability, we would detect remarkably little genetic marital assortment on this dimension.

12. It is, of course, possible that a wife with depressive genes is attracted to a husband who is depressive himself (or vice versa), but when we observe that there appears to be no average similarity between the spouses on their depression polygenic indices, this seems highly unlikely. I say highly unlikely, rather than impossible, because there could be a world where genetically depression-prone women are attracted to environmentally depression-prone men but not genetically prone men. But it seems doubtful that prospective spouses would be able to make such nuanced etiological distinctions when selecting a life partner. In theory, we could use this method to identify social genetic effects on any number of outcomes just by factoring out the polygenic index of the person and then seeing the effect of the spouse. But the extent to which the individual's PGI is only picking up part of the gene-outcome relationship, to the extent that there is nonrandom mating on the trait in question, the spouse-PGI effect may be proxying the original person's unmeasured genetic effects. The (almost) zero correlation of PGIs within the dyad makes that unlikely and thus allows for estimation.

13. T. Liu, J. Qi, and D. Conley, "Using Genetics to Study Spousal Social Contagion" (working paper, Department of Sociology, Princeton University, 2024).

14. While individual (possibly genetic) differences may cause people to stop or to continue through all the steps necessary to derive pleasure from using marijuana, social influences clearly matter as well.

15. Counter to the expectations of parents, experimental studies have shown that susceptibility to peers peaks at age fourteen and then declines. So, thirteen- and fourteen-year-olds are the most influenced, and teens' ability to resist influences and stick to values that differ from those around them starts to slowly rise after that (probably too slowly for most parents' comfort levels).

16. Laurence Steinberg and Kathryn C. Monahan, "Age Differences in Resistance to Peer Influence," *Developmental Psychology* 43, no. 6 (2007): 1531–43.

17. Ramina Sotoudeh, Kathleen Mullan Harris, and Dalton Conley, "Effects of the Peer Metagenomic Environment on Smoking Behavior," *Proceedings of the National Academy of Sciences* 116, no. 33 (July 30, 2019): 16302–7.

18. In this approach, we followed in the footsteps of economist Carolyn Hoxby. Hoxby, "Peer Effects in the Classroom: Learning from Gender and Race Variation" (working paper 7867, National Bureau of Economic Research, August 2000).

19. To truly call something an indirect genetic effect, we need to be sure that

the "partners" are randomly assigned. That we did by leveraging the year-to-year variation in genotypes within schools (for the gradewide analysis). But we also need to know that the genotypes we are measuring are the right genotypes and that they are not merely stand-ins for the environments of their holders thanks to genetic nurture or population stratification. Smoking is one of the phenotypes, like education but less so, where the effect of the PGI seems to partly reflect genetic nurture or population stratification. Thus, in proper experimental design, we would not only have information about grade-mates' genotypes, we would also have access to the genotypes of everyone's parents. In this way, we could examine the peer influence not of the total genotype of grade-mates, but only that part of their genome that was randomly inherited from their parents.

Moreover, in the ideal experimental set up, we would not rely solely on the polygenic index for smoking as our measure of peer genotype. Indirect genetic effects for a given phenotype, say smoking, could be driven by entirely different genes than are the direct genetic effects. What if, hypothetically, having very high-anxiety schoolmates drove individuals to smoke? My stressed-out classmates cause me to become stressed out and need nicotine to calm my nerves. In this scenario, we should measure the effect on my smoking behavior of my peers' anxiety genes (which may or may not be related to their own smoking behavior). We would need to conduct a separate GWAS and construct a novel PGI for indirect smoking effects. And to do that, of course, we would need thousands or even millions of genotyped respondents (and their parents) organized by school and grade.

If this sounds like a high bar to definitively show the existence of indirect peer social genomic effects, it is. But given what we know from animal studies, they almost surely exist in humans and our findings reflect, at least in part, true social genomic effects.

Either way, in the absence of the perfect human experiment, we can still learn about purely social dynamics by examining the social genome. While performing this study, we noticed something that was intriguing: non-white students seemed to be influenced by all peers in their grade when it came to smoking, but white students were only influenced by their white peers. (Smoking is one of the few polygenic indices that predicts well across racial groups, making this analysis possible.) Our sample size was too small to detect whether this was truly a real difference or whether it was just due to random fluctuations in our data. But a future study with a bigger sample might deploy genetic information in order to know exactly who influences whom by gender, race, and any number of other factors.

As in the case of spousal effects on depression, the nature of peer effects is interesting from a purely social-scientific perspective. The social genome can be a powerful tool to understand the human environment. Whether or not the influences of others' genes on my behavior are purely genetic effects—in the sense of the genes acting within their bodies, which then affect me—or whether they are also picking up parental effects and other environmental influences, the social genome still solves some of the knotty problems that have plagued the social sciences as they have tackled peer influence. Over the next few years, thanks to genetic data, we will learn more about how and when peers matter—in domains much more consequential than the ice bucket challenge. We—parents, especially—may not like what we learn. After all, as much as we might like to, we cannot really blame our child's best friends for our child taking up a smoking habit. But learning the truth about social life makes us more able to react correctly—as parents or policymakers—even if those truths are sometimes hard to swallow.

20. However, peers with a highly protective genotype (i.e., in the bottom tenth in terms of likelihood to smoke) do not prevent others in the grade from smoking.

21. Yonwoo Jung, Lawrence S. Hsieh, Angela M. Lee, Zhifeng Zhou, Daniel Coman, Christopher J. Heath, Fahmeed Hyder et al., "An Epigenetic Mechanism Mediates Developmental Nicotine Effects on Neuronal Structure and Behavior," *Nature Neuroscience* 19, no. 7 (July 2016): 905–14.

22. That's only half the social genomic effect of the parents' genes. The genes that they did pass on, they also kept a copy of, and those genes not only have their direct effects in the bodies of the children but also the indirect, genetic nurture effects in the parents. So, actually, we can just calculate a total PGI for the parents and see what effect that has over and above the effect of the PGI the kids inherited in their own cells.

23. Augustine Kong, Gudmar Thorleifsson, Michael L. Frigge, Bjarni J. Vilhjalmsson, Alexander Young, Thorgeir E. Thorgeirsson, Stefania Benonisdottir, et al., "The Nature of Nurture: Effects of Parental Genotypes," *Science* 359, no. 6374 (January 26, 2018): 424–8.

24. If the genotypes of parents have a big effect, what about those of siblings? Here the evidence is mixed. Kong and his colleagues parted ways with deCODE, so they no longer had access to those exquisite data. Several other datasets in the world (called *trios*) have both parents and at least one child. But precious few have quads—genetic data on both parents as well as two (or more) offspring—which are what is required to properly assess sibling impacts. Remember: if we don't have the parents' genetic information, we can't treat the offspring as a random draw, an experiment of sorts. Other than the Framingham Heart Study mentioned previously, I am hard pressed to think of many other data sources that are publicly available to researchers that contain quads. One study in Norway has recently come online that

has a number of full families genotyped. Likewise, there is a subset in the Estonian Biobank. It is possible that Ancestry and 23andMe are sitting on goldmines of complete families genotyped, but alas, those are almost as inaccessible to the run-of-the-mill researcher (i.e., one who does not have connections to the companies) as the deCODE data are.

Much more common are datasets that have at least a subsample of siblings but which lack data on the parents. The UK Biobank has about 20,000 siblings among its approximately 550,000 individuals. The U.S.-based National Longitudinal Study of Adolescent to Adult Health has a couple thousand. The Wisconsin Longitudinal Survey also has a sibling design. The list of such scraps of data goes on. Even pooled together, they don't rival Iceland's deCODE, but they start to approach a large enough sample to be useful.

Even without parental data, siblings themselves are useful, since, as mentioned earlier, taking the difference in genotype between siblings is almost the same exercise as having the parents' genotypes in the model to make the offspring a random draw. That's because for any two siblings, the differences between their genotypes are randomly determined. If I have an A-A and my sister has C-C at a particular location on chromosome 11, we know that difference resulted from random chance when we were each conceived, not because we come from different ancestral backgrounds with different chances of getting A or C at that spot. Any locations where we are not different just drop out of the calculations. I say that this is "almost" as good as having parents because using my sister's genotype as a control of sorts for my own by counting how many As I have compared to her combines two effects: it is the effect of me having two As in this case, but also the effect of her having two Cs (on me). So, to the extent that siblings' genotypes matter for each other's outcomes, we are not getting only direct effects by running what are called *sibling difference* or *sibling fixed-effects models*; we are also getting the sibling genetic nurture effects thrown in there, muddying the waters.

But along the way, Kong discovered that given the highly constrained rules of genetics, he could reconstruct missing genotypes of family members not in the data through a process he dubbed *Mendelian imputation*. Taking the preceding example again, if I am A-A and my sister is C-C, we actually immediately know each of our parents' genotypes at that location. Both parents are A-C because both donated an A to me and a C to my sister. This is a relatively uncommon example, however. As it turns out, we can reconstruct the average or combined parental genotype of our mother and father, and that's good enough, as it turns out, to re-create the random draw of our individual genotype from the collective parental gene pool in which each of us arose. Like Kong's earlier paper on genetic nurture, this calculation (which in retrospect seems obvious but which nobody had thought of before), changed the playing field for analysis of direct genetic effects, since any dataset with siblings could be turned into one with the prior generation's genotypes also factored out. (It should be noted that this method can be applied in other ways as well; for instance, given one parent and one offspring, one can impute the other parent's genotype.)

When a member of Kong's original team—Alexander Young—performed this imputation on the siblings from the UK Biobank and then looked for effects of sibling genotype on each, he and his colleagues found none. The story is not over, however, since suggestive evidence (and common sense) says that while the average sibling reciprocal effect may indeed be nil, that zero might obscure specific effects. Perhaps older siblings affect younger ones and not the reverse? Perhaps only siblings close in age affect each other? Perhaps only siblings of the same sex—or of one sex in particular—do so? Evidence on sibling effects from social science that uses natural experiments has found strong older-to-young spillover effects, so with enough data, we would probably find them with respect to genes, too.

More recently, the genetic nurture effects of parents have been called into question. Having parental genotype information ensures that the genetic effects of the next generation are indeed "true" direct genetic effects. But what Kong and others had been calling genetic nurture might not actually be the effects of the parents' genes acting on their behavior, but rather population stratification at the parental level. By studying nuclear families, we had conducted the ultimate natural experiment on the children, given the random assortment of their alleles. But we had no such experiment for the genetic nurture effects of parents on their offspring. Parental genotypes captured not only the effects of their genes in their bodies, but also all the assortative mating, population structure, and associated environments.

To say the genes of our parents are directly affecting us through the environment, we need data on grandparents, or at least on uncles and aunts so that we can impute the grandparents, rendering the parental genes the result of a Mendelian experiment. Very few datasets have such an extended family structure. The Norwegian dataset, MoBa, has some aunts and uncles, and when analysts used sibling differences at the parental level on these data, they found no pure genetic nurture effects. But their sample size was small and thus their analysis is hardly the last word. When a team led by Alexander Young used a better approach—imputing grandparents' genomes to create a three-generational data structure in three different datasets that we pooled—we did find some true genetic nurture effects. Sample sizes were still small, so for some phenotypes, we couldn't generate significant results, but going forward with larger samples, we should see a signature of genetic nurture for any phenotype we know to have a notable effect on family environment ("C" in the ACE twin models).

deCODE, with its oodles of Icelandic data, is reportedly about to produce unbiased genetic nurture effects of not just parents, but uncles and aunts as well. Stay tuned. Regardless of what turns out to be the right estimates of genetic nurture effects, within-family estimates of the "direct" genetic effects of one's own genes on some outcomes turned out to be smaller than the original GWASs had suggested. Factoring out both population stratification and genetic nurture set the football back not ten but twenty yards in terms of accounting for the heritability of outcomes.

For now, it is worth noting that for some uses of genetic data, it is critically important to factor out population stratification as well as genetic nurture, and for some applications, we shouldn't be troubled by a messier polygenic index that rolls into it some population stratification, assortative mating, and genetic nurture effects. If we want to identify who is at risk for developing hypertension or high cholesterol, we want a measure that simply predicts as well as possible. If that prediction happens to include a mixture of genetic nurture and direct genetic influences, then so be it. We don't care what the exact mechanisms are. There are use cases in social science as well that don't require a "pure" direct genetic effect measure. In the peer smoking case, where we use genes to measure the influences of peers, we are not all that concerned with whether our polygenic indices are picking up genetic nurture or population structure because the important aspect of the genotypes in this case is that they are not changed by our peers' behavior.

In other cases, however, getting the pure direct genetic effect is critically important. If you had to pick between two frozen embryos to minimize your child's risk of diabetes, you would want the effects to be direct genetic effects since the genetic nurture, assortative mating, and population structure influences would be the same since you would be implanting that embryo in the same womb, in the same household, and so on, no matter which you picked. Likewise, when it comes to doing science with polygenic indices, sometimes we want direct effects. When we talk about how parents respond to children's genotypes, for example, we very clearly want to isolate the genetic effects as they are expressed only in the child, as to keep the parents' own prior tendencies out of the picture. Thus, we need the magic of meiotic recombination to leverage only the random chance of which alleles a given kid happened to receive in the great genetic deck shuffle.

25. Dalton Conley, Benjamin W. Domingue, David Cesarini, Christopher Dawes, Cornelius Rietveld, and Jason D. Boardman, "Is the Effect of Parental Education on Offspring Biased or Moderated by Genotype?" *Sociological Science* 2, no. 6 (2015): 82–105.

26. The potential presence of multiplicative interaction effects also raises a thorny issue for the original Kong et al. study design in determining indirect, genetic nurture effects. Namely, if the effect of offspring genotype on offspring outcomes depends on parental genotype (and vice versa) then simply running a model that specifies independent effects of each genotype in the family on ego (i.e., the focal respondent) leads to misspecification (i.e., wrong estimates), since we would need a model that has higher-order interaction effects as well.

27. Though it should be noted that since ACE twin models leverage siblings, they do not suffer from confounding by genetic nurture (or population stratification—i.e., chopsticks genes).

Chapter 6

1. Madlaina Boillat, Pierre-Mehdi Hammoudi, Sunil Kumar Dogga, Stéphane Pagès, Maged Goubran, Ivan Rodriguez, and Dominique Soldati-Favre, "Neuroinflammation-Associated Aspecific Manipulation of Mouse Predator Fear by *Toxoplasma gondii*," *Cell Reports* 30, no. 2 (January 2020): 320–34.

2. In animal biology, these processes of altering one's environment and/or moving to a new environment are called *niche construction*.

3. "Joe Biden Should Run against the Ivy League," *The Economist*, July 12, 2023.

4. We must, of course, account for the fact that the education PGI does not pick up all the effect of genetics for the first question about the value-add of college. Moreover, for the second question, we need to consider not only this fact, but also that it picks up effects that, though they are pre-college (and thus not affecting the answer to the first question), may be environmental in nature, unless we use a PGI based on within-family GWAS (which then explains less variance).

5. Again, however, we are not factoring out all genetics with the PGI, so that may explain some of the difference.

6. This is a more modest effect than what recent research has found about going to an "Ivy Plus" school (the Ivy League plus MIT, Stanford, and the University of Chicago). When comparing students who "randomly" got off the wait list to those who didn't and ended up attending a flagship state university, those who lucked out had a 60 percent higher chance of ending up in the top 1 percent of the income distribution—though across the rest of the income distribution the effects were much more modest. Our analysis didn't focus specifically on the tippy top of the college pyramid nor on the 1 percent of the income distribution. See: Raj Chetty, David J. Deming, and John Friedman, "Diversifying Society's Leaders? The Determinants and Causal Effects of Admission to Highly Selective

Private Colleges," (working paper 31492, National Bureau of Economic Research, July 2023).

7. Many caveats to this analysis stem from the fundamental problems of genes selecting environments. For one, both PGIs only capture a portion of the genetic influence on education and income, respectively. So, it's possible that "college" is acting as a proxy for the unmeasured part of our genome and is not, in fact, having the actual effect we think it is from the analysis. Further analysis, however, casts doubt on that possibility. We can test this possibility by turning up or down the amount of the total genetic effect the PGI is capturing and seeing if the results change. This sort of exercise is called *simulation-extrapolation* or *SIMEX*. Since you cannot reduce measurement error to be less than what you've got already, SIMEX does the opposite: the procedure adds doses of measurement error to your estimator (the PGI in this case) and then plots what happens to the coefficient of interest (college or college selectivity in this case). We can plot a line or curve of what happens when noise increases, and then we can use that plot to extrapolate what would happen if we lowered the measurement error to zero. The results, for the most part, don't change. Another caveat is that by using the standard PGI rather than a within-family PGI, we are not just getting direct genetic effects but also indirect ones, the impact of assortative mating, and any environment that is correlated with the genes. But for the purposes of knowing whether college has a treatment effect, we actually want all those pre-treatment factors in there. Specifically, we want to factor out the effect of our parents' genes as well as ours and whatever childhood environments may be correlated with our genes, too. Moreover, a recent analysis by Akimova *et al.* (Evelina Akimova, Stefania Benonisdottir, Xuejie Ding, Melinda Mills, Shuang Song, Ramina Sotoudeh, Felix Tropf, Tobias Wolfram [2024] Quantifying the role of vertical pleiotropy in genome-wide-association studies. *European Social Science Genetic Network Conference.*) uses an exogenous (i.e., unrelated to genes) measure of schooling (the instrumented schooling level) based on a schooling reform in Britain as a control variable in a conditional GWAS. Without the control for education level, the common SNP heritability of income was about 20 percent. Controlling for education, it dropped to below 10 percent. Given some assumptions (most notably that the local average treatment effect [LATE] estimated by the policy change is generalizable to other levels of schooling, such as college), this tells us that in the UK, most of the effect of genes on income goes through education. Furthermore, from this analysis, we cannot ascertain whether the environmental treatment of "college attendance" is experienced differently by those with one set of genes (i.e., a high PGI) or another (i.e., a low PGI). It could be that the big bang for the tuition buck is obtained by people who have a high

education PGI and are better able to take advantage of the opportunities college offers. Or, rather, the opposite dynamic could hold. It could be the case that the high PGI kids are more likely to succeed through other routes if college is blocked (like it was for many groups in society for many decades) and that for the low PGI graduates, the diploma (or the college experience) has the greatest impact on later wages.

8. Of course, to the extent that people don't move because of their genes, that's also genetic nurture and passive GE; so even if it was our great-great-grandparents' genes that drove the family to the U.S., the fact that the subsequent generations stayed here might also be related to genes.

9. Rosalyn Price-Waldman and Mary Caswell Stoddard, "Avian Coloration Genetics: Recent Advances and Emerging Questions," *Journal of Heredity* 112, no. 5 (July 2021): 395–416.

10. Sometimes it's not so easy to distinguish between forms of GE correlation. The timbre, pitch, prosody, and complexity of a male songbird's call are also affected by his genotype. Singing is something active he's doing—so it has elements of active GE correlation; yet unlike the earthworm who tirelessly tills the soil, it is the response of another organism—the potential mate or competitor—that changes the environment. The *t. gondii* story is also an example of both active and evocative GE correlation. (See next paragraph in the main text for a definition of evocative GE.) It's active in that the parasite works to change the brain chemistry of the mouse by infecting it with cysts. But it's also evocative in that the response of the cat is what really matters—that it eats the mouse. All three genres of GE correlation may even be at work in a single case: dogs with sociable genotypes—that is, ones that like to be petted, to play with humans (or other dogs), and are all-around cheerful—are not only born to canine parents that provide a sociable environment (by already haven gotten themselves adopted into a human family, for instance); they seek more human affection by, say, nuzzling up against their human roommates (active gene-environment correlation); and their genetically determined looks and personalities evoke more social interaction in response (they receive more affection than a dog without that genotype would). They get, in other words, a triple dose of environmental inputs in response to their genes.

11. Certainly, nonhuman animals might also respond to humans' genetics, but I would call this social, too. I cannot think of any examples where the nonbiotic environment responds to a person's genotype. As much as we may think the sun only shines on us, the weather, air quality, and so on do not respond to us last I checked. At least not individually and in terms of genetic differences. Collectively, the environment responds to human activity, so much so that some now call the present geological epoch the Anthropocene—so

named because humans are recognized as having significantly altered the global environment. Some of this might be active, but the biome and climate are complex systems, so they "respond" in unpredictable ways.

12. Those outer layers can also have direct impacts on the child—not all socialization is mediated through the primary caregiver. But for the most part, each concentric circle impacts the one right inside it. Urie Bronfenbrenner, *The Ecology of Human Development: Experiments by Nature and Design* (Cambridge, MA: Harvard University Press, 1979).

13. Bronfenbrenner didn't think of the child as a completely passive agent, but his model didn't account for how very young children were already preprogrammed by their genes to have differing desires and capabilities.

14. We used the mother's behavior at age six months as a sort of negative control or placebo test. Our logic was that whether your kid is a genius or the opposite of a genius, there is relatively little that he can do at six months to convince you of that. If there were any effects of child genotype on parenting at six months, we thought it would be a sign that something else was seeping into our airtight analysis. Luckily, there was no effect of child genotype at six months.

15. The fact that the data come from the mother's survey responses rather than direct observation might also be relevant. A mother might "remember" more good parenting for a child who has turned out to be smart than for a child who has turned out to be less smart. However, mitigating against this possibility is the fact that parents were asked in real time about their activities before any record of academic success was evident.

16. Anna Sanz-de-Galdeano and Anastasia Terskaya, "Sibling Differences in Genetic Propensity for Education: How Do Parents React?," *Review of Economics and Statistics* (April 12, 2023): 1–44.

17. In this study, the exact language of questions asked of the parents about activities they spent time doing with their kids was as follows: "i) In the past 4 weeks went to a movie, play, museum, concert, or sports event with the mother (father); ii) In the past 4 weeks had a talk about a personal problem were having with the mother (father); iii) In the past 4 weeks talked about school work or grades with the mother (father); iv) In the past 4 weeks worked on a project for school with the mother (father); and v) In the past 4 weeks talked about other things were doing in school with the mother (father)."

18. Thomas Laidley, Benjamin Domingue, Piyapat Sinsub, Kathleen Mullan Harris, and Dalton Conley, "New Evidence of Skin Color Bias and Health Outcomes Using Sibling Difference Models: A Research Note," *Demography* 56, no. 2 (January 9, 2019): 753–62.

19. Though the distribution on skin tone genes is obviously shifted in non-Black populations, there is still variation that would likely have an effect if the mechanism were biological and not social. Another, perhaps better, test

would be to show that the skin tone genes don't have any effect on blood pressure in a different culture, where there is no pigmentocracy.

20. Further investigation found that it's not a simple story of evocative GE. It turned out that it wasn't their adult height that matters. It was their height in adolescence. If two six-foot-three-inch men were compared, one of whom was six feet tall by age sixteen and the other who was shorter at age sixteen but caught up later, the former man would, on average, get paid more. The researchers who conducted this clever study speculated that it was a matter of height-induced social confidence during adolescence that translated to a certain way of being—more self-assured, charming, secure, daring, and so on—in adulthood that employers were rationally responding to. It wasn't the employers who were favoring people who are genetically taller—since this example involves two men who are the same height in adulthood. Teenagers are the ones who are irrationally impressed by height, and that feeds back into the psychological state of the people doing the evoking. Teens who are taller end up more confident than those who are shorter, regardless of their final, adult stature. That confidence translates into more economic success later in life—so irrational evocative GE during adolescence may be rational evocative GE by employers (reacting to personality, not height) in adulthood. Or it could be that irrational evocative GE during adolescence based on height genetics leads to active GE correlation if the mechanism by which tall people tend to end up richer is by becoming more entrepreneurial, assertive, or risk-seeking, for instance.

21. It's particularly hard to fight our internal biases when there is a statistical truth to them. For instance, overweight and obese people generally suffer an earnings penalty. At the same time, thin people, on average, are in fact *slightly* more productive workers than overweight workers. That slight productivity difference may be related to the health effects that often tag along with being overweight—at least one study shows increased absenteeism and presentism (when someone comes to work sick) as the main explanation for lower productivity among obese workers. Or the slight penalty might be more directly related to the ability to perform physical tasks in a job. It might even be the case that impulsivity, which relates to eating habits, is also problematic for many jobs. Finally, it could be that the lousy treatment that heavier people get at work and in other domains of life impacts their self-esteem negatively, and this, in turn, results in a productivity difference.

But the wage gaps by body size are larger than what they should be based on all available data on productivity. That is, the difference in productivity (or healthcare costs) between overweight and normal-weight people is slight, but the differences in wages are much larger since employers may act as if this difference in productivity is much bigger than it is. That is, they may irrationally hold back heavy applicants and employees in the face of data suggesting that, based on productivity, they should be compensated more. To the extent that body mass index is influenced by genes—and estimates suggest that it is somewhere between a third and two-thirds genetic in origin—and that maximum productivity is the employer's true goal, then this is an example of an irrational, evocative gene-environment relationship.

There is additional evidence that the wage gap by body weight is at least partially a case of irrational evocative GE. Namely, when we look more closely, we discover some baldly irrational patterns of evocation with respect to body mass index. Unlike other demographic groups, overweight Black men enjoy a wage premium compared to their non-overweight counterparts. It's doubtful that this is a rational response—that, for instance, the genes for high BMI would cause lower productivity in Black women, white women, and white men but cause higher productivity in Black men. The explanation for the bonus for overweight Black men is that being overweight may mitigate some fearful reactions and stereotypes about Black men, since obese men may be seen as less "threatening." Moreover, the penalty is the strongest for women, which suggests that it has to do with irrational associations rather than anything related to productivity. Research shows that obese women are viewed much more negatively than their male counterparts, who are often seen as "jolly fat men."

Thus, we can pretty safely conclude that at least this aspect of looks-based discrimination represents doubly irrational evocation: it's probably not caused by the behavior of the individual in a given interaction (as evidenced by the online resume experiments), and it's probably not a statistical best guess on the part of the evoked (an interpretation supported by the experiments but also by the fact that the wage penalty for being overweight varies by demographic subgroup).

22. Under the assumption that there are not other sex-specific pleiotropic effects of those genes on wages through, say, aggression.

23. As far as I know, though some extreme environmental conditions like copper poisoning can affect eye color by generating dark circles called Kayser–Fleischer rings. Glaucoma medication can also change eye color.

24. The same dynamic happens with social attitudes such as conservatism/liberalism or religiosity. At age eight, social attitudes are 80 percent environmental and only 20 percent in our genes. But by age thirty-five, the genetic component has doubled. Unlike the case of cognitive ability, environment still contributes the lion's share—60 percent—of variation in social attitudes even in adulthood, but genes have learned to express themselves more forcefully as they have time to interact with the world. Our social attitudes are formed by our interactions with the people around us, the news, and

direct observation of the social world. As is the case for cognitive ability, environmental feedback is key; but in the case of social attitudes, the environments "chosen" by our genes burrow further into our ideological bunkers and receive reinforcement. It's not environment in the sense of random events happening—that gets less important as we age—it's environment in terms of the niche construction our genes create for us through selective social interaction. Active and evocative gene-environment correlation both play a role in the intelligence and social attitudes examples. But a counterexample is provided by attention deficit hyperactivity disorder (ADHD). ADHD, evidently, is more like an edifice than a machine learning bot. As early as age five, two thirds of the differences in attention deficit hyperactivity behavior is explained by genes, a figure that remains stable through childhood up to age eighteen. ADHD is not something that our genes hone and refine through collecting data out in the social world. I think this is because you don't learn ADHD; to become hyperactive or impulsive, you don't need to be taught. If you have impulses to jump up from your seat or blurt out the answer without raising your hand, the social response you get generally does not encourage you to further that behavior but does encourage you to hone your impulsiveness. If anything, the feedback you get is negative and seeks to restrain the genetic expression of that impulsivity or hyperactivity.

25. However, we can likely say that passive gene-environment correlations probably don't play as big a role because the power of our parents' choices (that stem from the genes they have and share with us, creating the passive GE correlation) wane as we age, since the common family environmental element ebbs.

There are several possible alternate explanations for rising heritabilities of traits as we age. First, genes could take their time in unfolding in our development. The genes for male pattern baldness wait patiently during a man's childhood and only manifest in adulthood—pretty much regardless of the environments the boy or the man experience. Ditto for gray hair and Huntington's disease, as mentioned earlier. Likewise, the genes that regulate the morphology of secondary sex characteristics—that is, pubic hair, voice changes, breast development, and so on—do not get activated until we are flooded with hormones during puberty. So, if we tried to measure something like the pitch of a male voice early in life, we would find it displays a low heritability, but if we wait until the teen years, we might find that genes explain a lot of the variation. Some scholars have offered this as an explanation: brain morphology keeps developing until our mid-twenties, and thus we don't see the full fruition of the genetic architecture until then (making it analogous to height).

A second possibility is that the differences in environments just wane as we age, making the influence of genes seem stronger in terms of explaining differences between people. Heritability is calculated as a proportion. It's the amount of variation in genes over the total variation in a phenotype in a population. If relevant environmental differences that individuals experience (not on account of their genes but unrelated to them) simply fall away as we get older, then it would reduce the total variation in the outcome, making it seem like genetics becomes more important, when all that happens is that the influence of non–genetically influenced environments decreases, leaving a bigger share of the variation to be explained by genetics. But there is no objective way to measure this. We could pick a bunch of indicators of environmental factors that we think matter and see if there is more variation in those measures as we age. However, we would not know whether these were the important indicators for the outcome—say, cognitive ability—about which we care. Moreover, when we think about variation in quality of preschools versus quality of high schools, or differences among parents in how they treat their one-year-olds versus how they deal with teenagers, it would seem that, if anything, environmental differences grow with age, they don't decrease.

Finally, even if we knew the magic X-factor in the environment that was the key aspect of nurture for the trait, and even if we observed that variation in the population increased or decreased by the age of a child for that X-factor, we would still not know whether that increased or decreased was due to random changes in the distribution of the X-factor (like a war) or due to the creation of different levels of X-factor by the genes of children who cultivate X to differing degrees. That's all to say that there is strong (but not wholly definitive) evidence that for certain traits, like political beliefs and cognitive ability, increasing heritability is due to the recursive loop between genes and the environment, that a growing child with a particular genetic predisposition seeks out the environments that reinforce and nurture that predilection. In general, the more freedom we enjoy to make our own, genetically influenced choices, the more our innate differences will lead to different outcomes, and the amount of freedom we have depends strongly on age, culture, and political system.

Chapter 7

1. Wesley Abney, "Live from Washington, It's Lottery Night 1969!," HistoryNet, November 25, 2009.
2. Abney, "Live from Washington."
3. Each twin is part of the environment for its sibling, by hogging or sharing the

placenta—thus showing that the social genome rears its head, so to speak, even before we are born (not to mention parental genetic nurture in the womb as well).

4. Technically, we can talk about average treatment effects of the environment summed across all genotypes and, likewise, average effects of genotypes across all environments, but that obscures the rich variation that is only visible when we consider both simultaneously.

5. MAOA is a gene that codes for the enzyme, monoamine oxidase. This enzyme, in turn, breaks down neurotransmitters when needed (after they've done their job). But if it's too active, then mental health problems can ensue—or, at least, that's the theory behind monoamine oxidase inhibitor drugs, one of the go-to classes of antidepressants before the discovery of Prozac. Avshalom Caspi, Joseph McClay, Terrie E. Moffitt, Jonathan Mill, Judy Martin, Ian W. Craig, Alan Taylor, and Richie Poulton, "Role of Genotype in the Cycle of Violence in Maltreated Children," *Science* 297, no. 5582 (2002): 851–54.

6. In 2003, Caspi and colleagues followed up this study with another *Science* paper that showed the exact same dynamic with respect to depression and the serotonin transporter gene (*5-HTP*). People with a short version of the gene who experienced childhood abuse or neglect were at higher risk for depression. But people without that risky allele experienced no increased risk due to abuse or neglect. And people with either the risky or non-risky gene who didn't get abused or neglected also had a lower rate of depression. That is, it took both the risky genotype and the less-than-ideal childhood environment to manifest the trait. But since genes for depression are likely also genes for morose parenting, it's a classic case of passive gene-environment connection, and we cannot say whether it is really a case of a risky gene being "activated" by an environment or, again, just unmeasured genes in the child that are shared by the parents (as indicated by their abuse/neglect) and raise the genetic risk alongside the measured serotonin transporter gene. See: Avshalom Caspi, Karen Sugden, Terrie E. Moffitt, Alan Taylor, Ian W. Craig, HonaLee Harrington, Joseph McClay et al., "Influence of Life Stress on Depression: Moderation by a Polymorphism in the 5-HTT Gene," *Science* 301, no. 5631 (2003): 386–89.

7. Even in the simple case of wanting to know the average net effect of attending college, we suffered from the problem that the PGI we were using to factor out genetics didn't pick up the entire genetic influence on college (or wages). But we were able to use statistics to extrapolate how much our results would change if we did measure the entire effect of genes. By using SIMEX (the simulation-extrapolation technique) we could simulate more measurement error and then extrapolate back to the case of what would happen if

we had no measurement error (i.e., picked up all the genetic effect). This got us to the full common SNP effect, but it didn't capture the effects of rare variants. It did give us a pretty good handle on the situation, nonetheless. Likewise, Akimova et al.'s work that uses an exogenous source of education as a control in a conditional GWAS for income also shows that education is a critical pathway for genes to realize income. (Evelina Akimova, Stefania Benonisdottir, Xuejie Ding, Melinda Mills, Shuang Song, Ramina Sotoudeh, Felix Tropf, Tobias Wolfram [2024] Quantifying the role of vertical pleiotropy in genome-wide-association studies. *European Social Science Genetic Network Conference.*)

8. Norman Hearst, Thomas B. Newman, and Stephen B. Hulley, "Delayed Effects of the Military Draft on Mortality," *New England Journal of Medicine* 314, no. 10 (March 6, 1986): 620–24.

9. Peter Dizikes, "The Natural Experimenter," *MIT Technology Review,* January 2, 2013.

10. Joshua D. Angrist, "Lifetime Earnings and the Vietnam Era Draft Lottery: Evidence from Social Security Administrative Records," *American Economic Review* 80, no. 3 (June 1990): 313–36.

11. It turned out that even the income differences among whites observed in the 1980s faded away in the subsequent decades, according to a paper cowritten by Josh Angrist. Joshua D. Angrist, Stacey H. Chen, and Jae Song, "Long-Term Consequences of Vietnam-Era Conscription: New Estimates Using Social Security Data," *American Economic Review* 101, no. 3 (May 2011): 334–38.

12. Again, some folks argued that the experiment was not perfect. For instance, James Heckman, another Nobel Laureate, argued that employers may have been unwilling to hire those men with "bad" draft numbers since they might invest in training them only to see them whisked off to Vietnam. I am not persuaded by this critique, however, because when Angrist ran his analysis on the 1973 cohort, there was no effect on wages for men who got "high" or "low" numbers but didn't end up needing to be drafted. This is a nice placebo test, since employers presumably would have discriminated by draft numbers for this 1973 group, too, before it was known that they wouldn't be inducted. Of course, employers might have realized that, with the drawdown in forces, a trainee getting drafted might have been getting rarer. Overall, however, I think the evidence is that Angrist's conclusions are solid.

13. Unlike previous draft lotteries for World War II, the Vietnam lotteries arrived at a Goldilocks moment in the history of human science. They began just when the systematic collection of data in durable formats (like punch cards and reel-to-reel tapes) had taken root, but before social and behavioral scientists became so enamored with field experiments that excessive efforts to study them degraded their "naturalness."

14. Stephen B. Billings, David J. Deming, and Jonah Rockoff, "School Segrega-tion, Educational Attainment, and Crime: Evidence from the End of Busing in Charlotte-Mecklenburg," *Quarterly Journal of Economics* 129, no. 1 (February 2014): 435–76.

15. Douglas Almond, Lena Edlund, and Mårten Palme, "Chernobyl's Sub-clinical Legacy: Prenatal Exposure to Radioactive Fallout and School Out-comes in Sweden," *Quarterly Journal of Economics* 124, no. 4 (November 2009): 1729–72.

16. Marshall Burke, Solomon M. Hsiang, and Edward Miguel, "Climate and Conflict," *Annual Review of Economics* 7, no. 1 (August 2015): 577–617; Nina von Uexkull, Mohai Croicu, Hanne Fjelde, and Halvard Buhaug, "Civil Conflict Sensitivity to Growing-Season Drought," *Proceedings of the National Academy of Sciences* 113, no. 44 (October 17, 2016): 12391–96.

17. Donald P. Green, and Daniel Winik, "Using Random Judge Assignments to Estimate the Effects of Incarceration and Probation on Recidivism among Drug Offenders," *Criminology* 48, no. 2 (May 27, 2010): 357–87; since Rama-dan follows a lunar calendar, it falls at different times of the year each year, providing some randomness to when people are exposed to it. See Douglas Almond and Bhashkar Mazumder, "Health Capital and the Prenatal Envi-ronment: The Effect of Ramadan Observance during Pregnancy," *American Economic Journal: Applied Economics* 3, no. 4 (October 2011): 56–85; Adriana Lleras-Muney, "The Relationship between Education and Adult Mortal-ity in the United States, *Review of Economic Studies* 72, no. 1 (January 2005): 189–221; David N. Figlio, "Boys Named Sue: Disruptive Children and Their Peers," *Education Finance and Policy* 2, no. 4 (Fall 2007): 376–94; Caroline M. Hoxby, "Does Competition among Public Schools Benefit Students and Taxpayers?," *American Economic Review* 90, no. 5 (December 2000):1209–38; Grant Miller and B. Piedad Urdinola, "Cyclicality, Mortality, and the Value of Time: The Case of Coffee Price Fluctuations and Child Survival in Colombia," *Journal of Political Economy* 118, no. 1 (February 2010): 113–55.

18. One problem was that WIC's rollout happened too fast, within a year, mak-ing it hard to parse differences. A second issue was that it improved nutrition not just in utero but afterward, so sustained nutritional influence could be what was having any salutary effect, not just via birthweight.

19. Dalton Conley and Emily Rauscher, "The Effect of Daughters on Parti-sanship and Social Attitudes toward Women," *Sociological Forum* 28, no. 4 (December 2013): 700–18.

20. Gordon C. McCord, Dalton Conley, and Jeffrey D. Sachs, "Malaria Ecology, Child Mortality and Fertility," *Economics and Human Biology* 24 (February 2017): 1–17.

21. Dalton Conley and Jennifer Heerwig, "The War at Home: Effects of

Vietnam-Era Military Service on Postwar Household Stability," *American Economic Review* 101, no. 3 (May 2011): 350–4; Dalton Conley and Jennifer Heerwig, "The Long-Term Effects of Military Conscription on Mortality: Estimates from the Vietnam-Era Draft Lottery, *Demography* 49, no. 3 (April 2012): 841–55; Tim Johnson and Dalton Conley, "Civilian Public Sector Employment as a Long-Run Outcome of Military Conscription," *Proceedings of the National Academy of Sciences* 116, no. 43 (October 8, 2019): 21456–62.

22. Romesh Vaitilingam, "Natural Experiments in Labour Economics and Beyond," *LSE Business Review*, October 25, 2021.

23. We lost some precision because some people who received the "treatment" (a draftable lottery number) didn't actually serve, and some people who received the "control" (an undrafted lottery number) volunteered anyway. But this noise was a worthwhile tradeoff given the problem of bias that the lottery solved.

24. A shout out to my astrophysicist wife for clueing me in to spectral analysis.

25. See, for example, J. Hunter Young, Yen-Pen C. Chang, James Dae-Ok Kim, Jean-Paul Chretien, Michael J. Klag, Michael A. Levine, Christopher B. Ruff, Nae-Yuh Wang, and Aravinda Chakravarti, "Differential Susceptibility to Hypertension Is Due to Selection during the Out-of-Africa Expansion," *PLoS Genetics* 1, no. 6 (December 2005): e82.

26. It doesn't even take something as dramatic as the fall of the Iron Curtain to induce such a gene-environment interaction effect. Even as recently as the first half of the twentieth century in the United States, the education PGI predicted poorly for half the population—the female half—because women were essentially barred from pursuing higher education. But when access to college improved for women, the education PGI started predicting better. See, e.g., P. Herd, J. Freese, K. Sicinski, B.W. Domingue, K. Mullan Harris, C. Wei, and R.M. Hauser. "Genes, Gender Inequality, and Educational Attainment," *American Sociological Review* 84, no. 6 (2019): 1069–1098.

27. It could be something about their home environments that generated the different responses, but given what we know about the impact of genes on reading and the fact that GxE interaction seems rife, my money would be on the genes explaining who reacted which way to the intervention.

28. Even identical twins have some small differences in their genomes.

Chapter 8

1. Economist Arthur Goldberger put it best: "if it were shown that a large proportion of the variance in eyesight were due to genetic causes, then the Royal Commission on the Distribution of Eyeglasses might as well pack up. And if it were shown that most of the variation in rainfall is due to natural causes, then the Royal Commission on the Distribution of Umbrellas

could pack up too." (Arthur S. Goldberger, "Heritability," *Economica* 46, no. 184 [November 1979]: 337.) Myopia is, in fact, about 70 percent heritable—roughly the same as cognitive ability, slightly less than height.

2. What is true, however, is that if a condition is predominantly genetic in its etiology, then if we address it, that fix only lasts one generation. Take myopia again: we can correct the vision of every nearsighted person in society for a matter of a few dollars each. But correcting individuals' vision does nothing to the vision of their offspring, thanks to the death of Lamarckism. We will have to keep providing glasses to nearsighted people who inherit myopia from their parents. Contrast that scenario to something like illiteracy in a poor, rural village. To the extent that parents are important agents of instruction to their children, if we teach people who had been illiterate to read, it is likely that their children will know how to read since the transmission is largely cultural and not genetic.

3. Another use of PGIs to diagnose environmental effects is to use them as instrumental variables in an approach called *Mendelian randomization (MR)*. I have not discussed this at length in the book since I am generally skeptical of the methods. Namely, for instrumental variables to be unbiased estimators, there can be no exclusion restriction violation (that is, no other pathways from the instrument to the outcome other than through the endogenous variable being instrumented). Given genetic correlations and how pleiotropic genes are, exclusion restrictions are almost always violated in MR. Thus, in order to know the effect of education on addiction, one might use the education PGI to instrument schooling and estimate its effect on substance use. But surely the education PGI has other pathways to addiction than through the poorly measured variable of years of schooling. The only cases in which I believe MR results are when there is a good placebo test. Say we used the coffee consumption PGI to see the effects of caffeine use on wages or on mental decline. We could show that among a group where, for cultural reasons, the caffeine pathway is blocked—say Mormons, who eschew coffee and caffeinated tea—the coffee PGI doesn't predict cognitive function or wages. Then, if we assume the putative exclusion restriction violations would be the same in the U.S. as a whole as among Mormons, we could rule them out and believe the IV estimates. This was the case for the alcohol study that I mentioned in Chapter 1, where the authors used Chinese women as a placebo group. For a lengthier discussion, please see Andrew McMartin and Dalton Conley, "Commentary: Mendelian Randomization and Education—Challenges Remain," *International Journal of Epidemiology* 49, no. 4 (August 2020): 1193–1206.

4. "N.Y. Preschool Starts DNA Testing for Admission," NPR, April 1, 2012.

5. Moreover, the Genetic Information Nondiscrimination Act of 2008 (GINA), which prevents genetic discrimination by employers and in the domain of healthcare, was meant to focus on single-gene diseases. But it's easy to see how big business might push for a revision of the law in an era where PGIs might be much more useful to them. The current Equal Employment Opportunity Commission genetic information discrimination regulations state, "An employer may never use genetic information to make an employment decision because genetic information is not relevant to an individual's current ability to work." But what happens when it is relevant to an individual's current ability to do their job and big-business legislators push to repeal GINA?

6. Though there has not been a lot of success with single-gene arguments, PGIs for mental illness may be used by the defense as exculpatory evidence.

7. This is not always a bad thing, since an efficient market grows the pie for everyone, even if the slices become more unequal.

8. Now "sniffers" can sample the air and get enough DNA from a room you were recently in to correctly identify you.

9. Though there is predictive information in each: your year and month of birth can predict some attitudes (the year you were born marks your "generation" and there are differences depending on month of birth since that affects health and schooling thanks to school cutoffs and the impact of being oldest or youngest in one's grade). And embedded in the first three digits of your Social Security Number (if issued between 1973 and 2011) is geographic information on the state in which you lived when it was issued that may be useful for targeting or prediction. Moreover, advertisers and others will not gain much traction from knowing your SSN, but they will be able to construct a profile of you from your various PGIs, maybe marketing diabetes drugs to you if you have a high genetic risk for metabolic disease or depression meds if that's your predominant polygenic risk factor. And so on.

10. John H. Evans, *The Human Gene Editing Debate* (Oxford University Press, 2020).

11. We also looked not just across institutions but across outcomes as well. We found that when the outcome being predicted involved medical issues, say schizophrenia, acceptability was highest (80 percent across all institutions). When it was more superficial, like with height (71 percent), the approval levels were slightly lower. (Approval regarding IQ fell in the middle, at 77 percent.)

12. This is particularly notable since—unknown to most people—wide use of polygenic prediction may engender yet more health inequities by race. Currently, PGIs predict the health outcomes of Americans with African ancestry less well than they do Americans of exclusively European descent. There are important scientific reasons for this difference like different levels

of genetic variation in the populations, but there are political ones as well. There are simply fewer studies with large numbers of Black subjects on which to train statistical models to calculate PGIs.

13. We didn't tell our subjects about the potential benefits or harms of polygenic embryo screening.

14. Of course, in vitro fertilization isn't the only context in which PGI scores could be used on embryos. Already, prenatal screening for major single-gene or chromosomal diseases (like Down's syndrome) has reduced the number of children born with major developmental challenges or shortened life expectancy. So, if a fetus is at higher risk for, say, schizophrenia or heart disease, how different is that than testing for Tay-Sachs? One important distinction is that genetic tests for single-gene diseases tend to be almost 100 percent accurate, while PGI prediction is noisy at best, meaning it is far from a perfect predictor. But even knowing that a high or low PGI score only marginally influences the risks for a given child, some parents may prefer to be safe than sorry. And what about selective abortion for non-disease traits? On the one hand, it may be the case that prenatal polygenic testing may lower rates of selective abortion since parents can screen before implanting. On the other hand, parents who get pregnant by accident the old-fashioned way and then want to make sure the fetus is a "winner" before bringing it to term, may elect to abort if the fetus isn't up to snuff.

One can assume that most Americans today would be horrified by the idea of aborting a fetus because it scored lower on the height or cognitive ability PGI than might be expected given the corresponding scores of the parents—just as vast majorities disapprove of sex-selective abortion (and some states have banned it). After all, selecting one embryo out of a dozen for implantation is quite different than terminating a natural pregnancy thanks to a score on a genetic index. But norms can change, of course, and it is easy to imagine a world where some parents, seeking every possible advantage for their offspring, wade into this territory, especially if others have already deployed selective abortion for disease avoidance. This is all to say, even a debate as entrenched in U.S. culture as the abortion debate will be touched by this new technology.

15. In her book *The Genetic Lottery*, psychologist Kathryn Paige Harden notes that three policy choices become available when we recognize the stratifying effects of genotypes: (1) current approaches, where we fly blind, which can exacerbate existing differences since those with advantageous genotypes will benefit more; (2) eugenic policies that privilege those who are already lucky in the genetic lottery; and (3) her preferred "anti-eugenic policy," which provides more resources to those who were unlucky in the genetic lottery. Harden argues that we don't want an anti-eugenic policy that holds more genetically advantaged people back from their potential; we want to provide stepstools to raise up those with disadvantaged genetic draws.

However, decades of political science and sociological scholarship have shown that progressive policy to aid the "truly disadvantaged," in Harvard Sociologist William Julius Wilson's words, has the best chance of being implemented, the longest lifespan and, relatedly, the broadest public support when it is not, in fact, means tested or otherwise targeted, but instead it is implemented as a public right of citizenship or something close to that—that is, an entitlement. A related observation by scholars has been that redistributive, Rawlsian-like policies tend to be implemented when the public perceives the cause of, say, poverty to be from completely external, recognizable forces that are orthogonal to individual characteristics (though it is unclear whether genes are seen as individual characteristics or luck) and/or when they are provided in response to a responsibility, onus, or demand asked for by the state of its citizens. Political scientist Theda Skocpol traces the origins of the New Deal welfare policy to the pensions provided to the widows of Civil War soldiers who gave their lives for the Union. It's no surprise we got the modern welfare state in the U.S. during the Great Depression. And we saw that same dynamic in play during the Covid-19 pandemic with respect to income support policy. The point here is that while, in theory, tailoring educational, health, or social policies to individuals to achieve maximum efficiency might, on the margins, be more efficient and save some wasted resources, there are political costs to further individualizing policies, which would probably be much more dramatic than even means-testing currently is. Such an effort to genetically "tailor" programs has the potential of weakening overall support for any redistributive or equalizing policies.

16. Michelle N. Meyer, Paul S. Appelbaum, Daniel J. Benjamin, Shawneequa L. Callier, Nathaniel Comfort, Dalton Conley, Jeremy Freese, et al., "Wrestling with Social and Behavioral Genomics: Risks, Potential Benefits, and Ethical Responsibility," *Hastings Center Report* 53, no. 1 (March 2023): S2–49.

17. Such policies end up being mandatory only for those "compliers" who genetically would not have taken the treatment absent being forced to. That is, even universal requirements do not, in essence, affect the always takers—that is, those who would have stayed in school anyway thanks to, say, advantageous home environments or genotypes.

18. See, for example, A. C. Heath, K. Berg, L. J. Eaves, M. H. Solaas, L. A. Corey, J. Sundet, P. Magnus, and W. E. Nance, "Education Policy and the Heritability of Educational Attainment," *Nature* 314 (April 25, 1985): 734–36.

19. James E. Cooke, "What Is Consciousness? Integrated Information vs. Inference," *Entropy (Basel)* 23, no. 8 (August 11, 2021): 1032.

INDEX